Allagash

A Journey Through Time on Maine's Legendary Wilderness Waterway

by Gil Gilpatrick

DEDICATION

**In memory of my father, Volney T. Gilpatrick Sr.
Known to Mom and his friends as Von, to us as Dad.**

ISBN 978-1-56523-487-1

Library of Congress Cataloging-in-Publication Data

Gilpatrick, Gil.

Allagash : a journey through time on Maine's legendary wilderness waterway / by Gil Gilpatrick.

p. cm.

Originally published: Skowhegan, Me. : G. Gilpatrick, c2003.

Includes bibliographical references.

ISBN 978-1-56523-487-1

1. Allagash River (Me.)--Description and travel. 2. Allagash River Region (Me.)--Description and travel. 3. Gilpatrick, Gil--Travel--Maine--Allagash River Region. 4. Allagash River Region (Me.)--History. 5. Historic sites--Maine--Allagash River Region. 6. Natural history--Maine--Allagash River Region. 7. Abenaki Indians--Maine--Allagash River Region--History. 8. Lumber trade--Maine--Allagash River Region--History. I. Title.

F27.A4G53 2010

974.1'25--dc22

2010029870

Printed in United States of America
First printing: August 2010

CONTENTS

INTRODUCTION **1**

PART ONE, ALLAGASH WILDERNESS WATERWAY **9**

- Prologue 11
- Chapter One, Put-Ins 15
- Chapter Two, Chamberlain Lake 35
- Chapter Three, Eagle Lake 41
- Chapter Four, Churchill Lake and Chase Rapids 51
- Chapter Five, Umsaskis and Long Lake 59
- Chapter Six, Round Pond 67
- Chapter Seven, Allagash Falls 73
- Chapter Eight, To Allagash Village 81

PART TWO, LUMBERING IN THE EARLY YEARS **85**

- Prologue, Bangor 87
- Chapter Nine, Dam It All 89
- Chapter Ten, Mushrat Murphy 99
- Chapter Eleven, The Tramway 111
- Chapter Twelve, Jill-Pokes in the Willywags 123
- Chapter Thirteen, The Eagle Lake & West Branch Railroad 141
- Chapter Fourteen, Big Jim's Crew 151
- Epilogue, Wood Harvesting Today 161

PART THREE, NATIVE AMERICANS **163**

- Prologue 165
- Chapter Fifteen, The Bear Clan 167
- Chapter Sixteen, Hawk-In-Flight's Moose 175
- Chapter Seventeen, The Burn 181
- Chapter Eighteen, Reunion With Grey Wolf 189
- Chapter Nineteen, Rogue Bear 197
- Chapter Twenty, The Goose Hunt 205
- Chapter Twenty-one, Running Bear's Journey 211
- Chapter Twenty-two, Winter Camp 217
- Epilogue 221

GLOSSARY **222**

BIBLIOGRAPHY **223**

ACKNOWLEDGMENTS

Without the help of some very dedicated individuals this book would simply be a reprint of an earlier edition and would have nothing new to offer. You will read about some of their exploits in Part One. I only wish there were room to fully credit them for all the work they have done for the Allagash Wilderness Waterway. First of all my thanks go to my wife, Dot, who not only reads and edits my work, but also patiently follows me around as I collect photographs and information. She is the one who makes it all worthwhile for me. My thanks to Terry Harper for making available to me his extensive research and the large number of historical photos he has collected. Thanks go to Tim and Sue Caverly for allowing me to use their photos and for keeping me informed on what is happening in the political arena that affects the Allagash. Finally, a thank-you to all of the folks in the Allagash Alliance Group for your work and dedication in keeping the wild in the Allagash Wilderness Waterway.

INTRODUCTION

In the thirty-plus years that I guided canoe parties through the Allagash I found my guests' interest about equally split between the natural attractions and the relics of bygone logging days. Surveys by the Bureau of Parks and Recreation show the tramway to be the number one area of interest. But of course, enjoyment of the natural attractions and beauty are cumulative, less tangible, and harder for most people to express.

Many people I have met who are making the trip, or have taken it, seem to have missed the overall concept of the dams, the tram and trains, and the other works that early lumbermen undertook to exploit the area's resources. They visit them, but don't fully understand their purpose. I hope my diagrams, photos, and descriptions help you have a better understanding of what those old-timers were up to. Their genius is absolutely amazing. Many of the schemes and methods of those early lumberers seem destructive today, but we must remember that it was a different time—a time when wilderness was something to be conquered, not preserved.

Part One of this book is meant to be a guide to the Allagash traveler. Like the other parts, it can be read piecemeal, taking in those parts that are relevant at the moment. It was also my intent to make it possible to make connections between where the reader is in Part One (either in the book or on the ground) and the history of that place in Part Two or Three. This to me is what brings history alive—the connecting of relics to their historical use. This is easily done with the logging history, because there are so many artifacts to see. Evidence of the Native American's use of the Allagash is far more subtle, but there nevertheless.

The descriptions of the rapids and other natural occurrences along the waterway are meant to be typical and not specific. There are guides to rivers, including the Allagash*, that offer far more detail than will be found in this book. However, my intent was to make the trip through Part One as interesting and relevant as possible. As valuable as the river guides are when specific information is needed, they are extremely boring to read for general information.

When modern Allagash visitors start to look around, they find evidence of white man's use that dates back to before Maine was made a state in 1820. If they are the curious type they ask, Why were these dams built? How did the men live in this wild country with no civilization nearby to support them? Why are old machine parts littering the shores of some of the lakes? How did they move the timber from here to the mills? Two giant locomotives rusting in the woods—amazing, why? How was such a far-flung operation supplied? And finally, what is going on today?

These are just some of the questions that plagued me during the early years I made trips through the Allagash region. Over time I have managed to accumulate many of the answers. Some, I admit, still defy me.

I found that there were huge gaps in the history of the region. The major events were mostly well documented—the building of the dams, the tramway, and the railroad—but in between were many years that were written about in very general terms, if at all, and in a context that could apply to any other region of the state. A chronological story of the area was going to be impossible to write without leaving large gaps.

I finally decided to make the lumbering history as readable and as interesting as possible by emulating the writers of historical novels. I filled in between the documented facts with a fictional account of what probably happened or, at least, what could have happened. Therefore the chapters more or less bounce from fact to fiction, but the fiction is written as close to the truth as I was able to determine. All of the incidents described really happened, though possibly not all in the Allagash.

The chapter entitled "Dam it All" is one hundred percent fact, at least as far I can ascertain. The various sources I used showed some variance in who got credit for what, but this isn't unusual. Different writers saw the various characters in different lights, so a hero to one was a villain to another. I tried to make my account as objective as possible, bearing in mind that these people lived and worked in a different era. What was considered progress then might

* Appalachian Mountain Club, *River Guide Maine*, second edition, 1991

be thought of as despoiling the environment today. One should try to judge these people in the light of their time, not by today's standards.

Mushrat Murphy is a fictional character. There were some Murphy lumber camps around Telos and Chamberlain, but there is no connection. I created him to show the way of life of the early woods workers. The way of life I depicted is based on fact. Most of my information on life in early lumber camps came from written sources, but some came from talking to people who had been there—including my father, who started his adult life as a cookee (cook's helper) in a lumber camp at age 13.

The story of the tramway turned the Allagash history into the 20th century. What an audacious undertaking! No need for fiction here. Think of it: A conveyor three-quarters of a mile long to carry saw logs, located a hundred miles or more from the nearest railhead or road. Was the tramway successful? Well, it worked and it was kept in operation for six years, but did it make any money for the stockholders? Everything I could find to read on the subject is positive, but between the lines they all seem to be defending it. Defending it from what or whom? These are questions I can't answer, but I suspect the project was obsolete, or nearly so, by the time it was put into operation, as the technology of the time brought forth better ways of moving logs out of the woods.

Jill-Pokes in the Willywags! The way it just rolls off the tongue, I had to use it for a chapter title, and it does fit the men it describes. The rivermen of Maine were every bit as colorful as the western cowboy; they just never got the press! The ballad by Holman Day at the end of that chapter makes that point. There are many parallels between the two. Both worked long, hard, thankless hours. Both honed the skills of their trade to a razor edge. Both took fierce pride in these skills and never missed the opportunity to show them off. There were exaggerations of their skills aplenty, too. For instance, the cowboy who could shoot an animal from so far away he had to put salt on the bullet so the meat wouldn't spoil before he got to it. And we shouldn't forget the river driver who could throw a cake of soap into the river and then walk on the bubbles. There were many audacious stories, and for sure neither the riverman nor the cowboy would walk across the street to deny any of them. The characters in Jill-Pokes are fictional, but again, the account of the life these men led is accurate and the events did or could have happened along the Allagash River. McGargle Rocks were named for a river driver who was drowned there, but I could find no more information than that, so the story is fiction, but the event fact.

The two giant oil-burning steam locomotives at the tramway are the biggest tourist attraction on the whole waterway, or at least the biggest man-made attraction. The attention is well deserved, for even with today's heavy machinery in the woods I imagine there would be considerable head-scratching if someone were to suddenly give the order to move the engines out of the woods.

In the early 1930s the railroad equipment was moved in with steam-and gas-powered Lombard Log Haulers in the dead of winter. It is as if the builders of this scheme of wood transport were trying to outdo the engineering feats of the tramway builders of a quarter century before. And outdo them they did. Their monuments still stand: the locomotives, the pulp cars, the graveled right-of-way (albeit tree-covered), and the trestle across Allagash Stream. If a company left such a mess in the woods today it would be in court in no time, but it is good these relics are here. They remind us of a time that is past, of a different era, of a different kind of thinking, and of a different set of values.

The story of the Allagash horseboat is half fact and half conjecture. Big Jim Fleury, Ma Tukey, and Alec Tidd all were real, and the events I describe really happened. The names of the rest of the crew were not available, and it is possible that the memories of the sources of the various facts have fused many river trips into one similar to that I described in "Big Jim's Crew." Whether Ma really did carry Little Alec off to the preacher I do not know, but the threat she made to do so was real. Maybe if there is a family of Tidds somewhere in Maine they can shed some light on this episode.

For me, the true fascination with history comes with being where it happened. I think this is why I have enjoyed researching and writing this part on lumbering. If you, the reader, enjoy it even a fraction as much as I enjoyed writing it, I have been successful!

Part Three of this book, about the Native Americans, has been pleasurable and yet frustrating to research and write. Proceeding from reference to reference, I found that I was dealing with the science of ethnology, the science of archeology, the historical writings of early explorers, and the legends and tribal memory of the people themselves.

The frustration came as I read one reference after another trying to form a clear picture of these amazing people. But no picture came. What one writer said as absolute fact, another would contradict. The problem, I finally decided, was that I was trying to write a historically accurate story about a people who had no written history! Realizing this, I made decisions based on the best information I could find, and the picture started to come together. When I found agreement in my reading I seized upon it and made sure to incorporate it in the story. When there were gaps or disagreement I chose the route that seemed most logical to me.

Sometimes the different sources and sciences answered the questions they created. While reading about the New England Natives in general I found frequent references to a class system in their government and society—a hierarchy, so to speak. But when I read of the Wabnaki in Maine I found no such class distinction. This seemed strange until I found a passage in one of Fannie Hardy Eckstrom's books that said there was strong evidence that our Maine Native Americans are more closely related to the tribes farther west (Fox, Shawnee) than to the tribes in nearby Massachusetts. Why this is, I do not know, but it does explain the strong difference in culture between the neighboring tribes.

The writings of the early white explorers were often contradictory and often left me scratching my head trying to decide which one to believe. A few of them tried to present a fair picture of the Native American and the way he lived, but too many others wrote as if to describe another form of animal life—a little more intelligent than the rest. What a shame this superior attitude had to exist, because it has robbed us of the only chance we had to see the Maine Wabnaki as he was before the white man changed him.

One early writer commented on an alleged Indian story he had heard. He decided that the story was of European origin and not Indian because it involved romantic love between a man and woman. "The Indians," he stated, "were not capable of romantic love as we know it." What nonsense! There were, no doubt, cases of arranged marriages among the Native people just as there were in many white societies, but to deny that a man and woman felt love for one another is ludicrous. No wonder there is so much confusion in the literature.

I grew up on stories and movies of cowboys and Indians. Like many others, I pictured the Native American woman (wrongly referred to as a squaw by white writers) as a witless drudge who was at the beck and call of her husband and had absolutely no say about the course of her life. I was pleased to find this picture was not true, at least not with the Wabnaki. The women did have a say in tribal government. Some of them became Sahgum, or chief, of their tribe. They lived healthy outdoor lives of hard work and were justly proud of their position in the family, clan, and tribe. Historic records show that many white women taken captive by the Wabnaki refused to return to their former pioneer life when given the opportunity. This says a lot about the status of the Wabnaki woman.

The misconceptions about the Wabnaki women stem from the fact that the women did the work the white observers were most likely to see and write about. They did all of the agricultural work and other chores around the village "while the men lazed around." But the men worked at more strenuous pursuits—hunting, fishing, warfare, and clearing the land.

One has only to imagine himself in the wilderness with nothing but his own mind and hands as tools to appreciate how hard the Wabnaki had to work to earn a living. Everything he had was made from the materials at hand. He made his tools to make his tools. But there was some division of labor and time for leisure and tribal celebrations. The mental picture some people have of lazy, no-account Indians comes from descriptions of a generation of these people who were caught between two cultures; their ancient and traditional one, and that of the white man, which they found nearly impossible to adopt. They just had no place to go and nothing to do.

Maybe I'm a romantic, but the legends of these people told me more about them than all of the archeological digs, ethnological theories, and historical records. These represented to me the voices of the people themselves speaking to us through the generations. I am no expert at interpreting these stories, but they are fascinating and certainly food for thought. They can at once entertain children, provide an explanation for a natural phenomenon, and provide a means of passing a culture from generation to generation.

Even with my lack of training in understanding these stories, I could see in some of them a direct tribal memory back to the ice age. And their cultural hero, Gluskabe, and the things he did are similar to stories about other notables who did some pretty miraculous things. Could they be the same people, some remembered through the written word and some by legend?

One of the ice-age stories is about a huge beaver that made a dam across the St. John River in what is now New Brunswick. The dam was so high that the water had to flow north to the St. Lawrence. Gluskabe went to correct this and created the reversing falls of the St. John when he broke the beaver's dam with his huge stone axe. As Gluskabe chased the beaver up the river he struck at it with his axe. The mud splashes caused by these blows created the islands in the St. John.

Anyone can attach a meaning to stories like this, but rivers reversing directions and the creation of islands have the ring of the ice age to them. These are the things that happened then, and scientists tell us that there were prehistoric beavers that weighed as much as six hundred pounds!

Religion was another area in which the Wabnaki was misunderstood. Because everything in nature was sacred to him, the early white men assumed he had many Gods. This is not true. The Wabnaki recognized one Great Spirit, or Manitou, who controlled all things. The red man did pray to different spirits, but this is understandable. After all, who would expect the great Manitou to concern himself with the day-to-day lives of each individual? When we think about it, is this practice so different from that of many of our modern religions?

Wabnaki technology never went beyond the use of stone tools, but they made science and art of this, leaving thousands of examples of their stone points, knives, scrapers, gouges, and axes. What a great amount of love, patience, and skill must have gone into them! And look at the decorations they used on tools, clothing, and implements. Is this the work of a cold, unloving people?

The birch canoe is said to be the most complex and intricate product of native mechanical genius in the north. Its design was already ancient when the first explorers reached these shores. The Penobscots credit their hero, Gluskabe, with teaching them how to build canoes. Some archeologists

Maine State Library collection.

A birch canoe. The white man found it so useful that all of the good birch bark was used, leading to the invention of the wood-canvas canoe.

suggest there is evidence that the birch bark canoe may have been invented right here in Maine. In any case, the birch canoe was used almost unchanged until the mid 1800s, when a Bangor craftsman invented the wood-canvas canoe. It is interesting that the reason for this change in materials was not because the wood-canvas canoe was considered superior, but that the supply of canoe birch was running out. The white man had used too much.

One thing the archeological findings tell us is how well-traveled the Wabnaki people were in their finely crafted canoes. Stone implements made of the felcites from the Munsungun mountains have been found throughout the lower Penobscot River valley and the entire waterway of the Allagash and St. John. That these people's lives revolved around these waterways is undeniable. No significant archeological site has ever been found that was not on or near a waterway.

The Native Americans lived with nature, bending with her ways, and so left few permanent marks upon the land. A fire-cracked rock here, a chip of chertz there is all that remains, and finding them requires luck or the trained eye of the archeologist. Once in a while an arrowhead or stone knife turns up for some lucky traveler, but most of these artifacts have been covered with the water raised by dams that have been in place since the middle of the 19th century.

The spirit of the Wabnaki is still there, though, and it is not difficult to feel. Sit at night on the lonely shore and listen to the call of the loon. Watch the giant moose dip its head beneath the water to reach the succulent water plants. Enjoy the crystal clear water, nearly as clean and unpolluted now as a thousand years ago. Lie at night and listen to the deer approaching your campsite, blowing with alarm but too curious to run. These are the legacy of the Native American, and they tell us much more about him than the few chips of stone he may have left behind.

This then is how I came by the material to write this story of one stone-age family's journey into and through the Allagash region almost a thousand years ago. The events are presented as what could have happened. No one knows what did happen. The beautiful Allagash was not an untrodden wilderness then any more than it is today. It was a region that gave life to a people who husbanded its resources for thousands of years. Let us hope that we use it as wisely.

I hope you enjoy this Twenty-first Century edition of Allagash.

Gil Gilpatrick

PART ONE

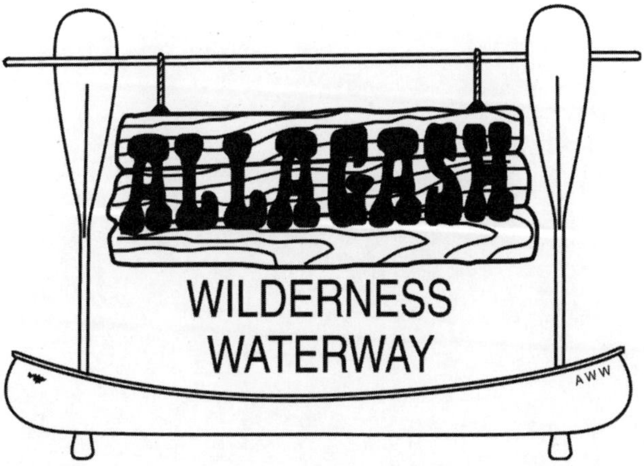

Allagash
Wilderness Waterway

PROLOGUE

In 1961 the United States Department of Interior Bureau of Outdoor Recreation proposed the establishment of an Allagash National Riverway with 150,000 acres of land surrounding the Allagash to be purchased by the federal government. This plan included the headwaters and the river itself, but also extended east to include Haymock Lake, Spider Lake, Pleasant Lake, Clear Lake, and the Musquacook chain of lakes. Hunting and fishing would be allowed, but timber cutting and mining would be banned. Access would be restricted to three points, and all motors, including motorized canoes and float planes, would be excluded. Only one logging road would be permitted to bridge the riverway.

The federal proposal was a shock to Maine residents, the legislature, and Allagash landowners. It implied that if a state-sponsored program was not offered immediately, the federal government would take over. The landowner's major concern was not the loss of 150,000 acres of timberland—that was less than 1 percent of the commercial forest land in northern and eastern Maine. Rather, they worried that once the foothold was gained the federal government would, in time, buy more land and further restrict lumbering operations. Also, they objected to having a long narrow strip split land holdings on the west from those in the east, with only one crossing over nearly one hundred miles.

In 1963 Maine created the Allagash River Authority and Advisory Committee to plan the protection of the Allagash region. In a letter to State Senator Edward P. Cyr, Senator Edmund S. Muskie stated that if the establishment of a Allagash wilderness area was to be meaningful it should "include the Allagash River and adjacent land areas as a contiguous and well defined entity irrevocably dedicated to its maintenance in a wilderness state." Then in 1964, Senator Muskie wrote to Forest Commissioner Austin Wilkins of the Allagash Authority that he and Secretary of Interior, Stewart L. Udall both felt "that the key issue...is the preservation of the riverway as a free-flowing stream in a primitive and, insofar as possible, unspoiled forest area. To be meaningful, such preservation must be in perpetuity."

In November 1966, Maine voters passed the Allagash Wilderness Waterway bill. The bill included a $1.5 million bond issue "to Develop the Maximum Wilderness Character of the Allagash Waterway." The bill was approved by 68 percent of Maine's voters.

Through the Allagash bond issue of 1966 the state acquired the land and water for the waterway. The government paid about three million dollars—half federal, half state funds—for the 22,760 acres of land and the 30,000 acres of water within the waterway's boundaries.

The Allagash Wilderness Waterway was born. Over the next ten or twelve years the number of visitors to the area climbed from about four thousand per year to nearly nine thousand. "How can you call it a wilderness with thousands of people crawling over it?" longtime opponents of the project complained. However, use of the area has leveled off, and through the determined efforts of the Bureau of Parks and Recreation, various watchdog groups and the daily vigilance of the rangers on the waterway, the wilderness character is being preserved. Thousands of people from all over the country have enjoyed this great wilderness experience. The people of Maine are to be commended for the foresight and good sense to preserve this beautiful area for now and for future generations.

For more than 30 years I guided parties through the Allagash Wilderness Waterway. Sometimes I made five or six trips a year. Although I never kept count, I have certainly made well over 100 trips. People often ask elders like myself what changes they have seen over the years. When I stop to think about it, I realize there have not been as many changes as one might expect. Oh, the people working there change from time to time, but the waterway itself has remained fairly stable in spite of the intrusions by mankind in several different ways.

Of course, the Allagash is not a true wilderness, being right in the middle of a working forest. As you read on through Part II of this book you will see that the Allagash was even less of a wilderness 100 to 150 years ago than it is now. If you define wilderness as "untouched by the hand of man," then even events in Part III would disqualify the area for wilderness classification, because even those Native Americans made subtle changes to enhance their lives.

Given that the Allagash is not a true wilderness, it is still important that the wilderness *character* be maintained. This wilderness character gives the visitor a feeling of wilderness without the inconvenience and danger of true wilderness travel. As travel through the north woods is relatively easy nowadays, with all of the private roads open to the public, various interest groups try to make changes, each of which could take away from, or add to, the wilderness character. These conflicts between various interest groups often elicit a lot of interest and many times hostility. Still, as these things play themselves out, the area remains a wonderful place to visit, spend some time outdoors, and learn a little about Maine's history.

During my years on the waterway, guests often asked me what interesting things happened on past trips. To come up with something to tell them, I often let my mind wander through the Allagash to places where something notable happened. At other times they would ask about a specific subject and that would jog my memory. I have treated you, the reader, like one of my guests and have recalled some of these stories. You will read several bear stories in the following chapters that I told when that subject was raised. They are sprinkled around this part of the book, coming up where they happened for the most part. I hope you find them interesting.

I often get calls, mail, and email asking for information about the Allagash. One of the most frequent questions is what campsites I recommend. That's a hard one. Even my wife, Dot, and I don't agree on some campsites, so how can I expect other people to like what I like? The reality is that almost all of the campsites have some quality that appeals to someone. Some people like a wide-open campsite with wide-open views, while others prefer a wooded

The typical Allagash campsite has a table with a fly pole over it, as seen here, a fireplace with a steel grate, and an outhouse. Some have camper-built furniture like the bench in the photo.

area with barely enough open space to set up tents. All of the campsites have the same facilities, so personal preference is key to selection. I have pointed out a couple of sites that I recommend you avoid, as they are just plain bad and fit only for emergency use when nothing else is available. Other than that, campsites are mentioned only as a matter of interest and not to recommend them. Maps are available that will point out all of the campsites and these should be referenced as you plan your itinerary and make your way through the waterway.

Descriptions of the rapids are not meant to be a white water guide by any means. They are only to make the reader aware of the white water that will be encountered. If you have the opportunity to learn a little about white-water canoeing before your Allagash visit you will find it a great help, especially in Chase Rapids. Even just a couple of hours with an experienced person will help you learn some basic white-water canoe strokes that will assist you in avoiding an upset and add to your enjoyment. If you are apprehensive about your first river trip, you might be wise to consider joining a guided group, where an experienced river guide will be there to teach the strokes and give advice along the way. A lot of the guests I had on trips over the years used the instruction I gave them to become enthusiastic canoeists and do trips on their own. I still hear from some of them from now and then.

The longest chapter in this part of the book is "Put-Ins." It is not that there are so many put-ins, but that some of them are trips themselves and so have been described in that chapter. By reading Put-Ins you should be able to customize just about any Allagash canoe trip to meet your needs, interest, and time. The remaining chapters in Part One are a mile-by-mile description of the waterway, with some anecdotes sprinkled here and there that I hope you find interesting and entertaining. Places that may be of interest are pointed out, and wherever possible. I have referenced those to other parts of the book. Although I don't intend to tell anyone how this book should be read, I hope you will find it interesting to look forward to Parts II and III as you make your way through the waterway. I have assembled a number of old photos just to compare then with now.

[CHAPTER ONE]

PUT-INS

Put-in is almost certainly a term exclusive to canoeing and related sports, I think. It means, of course, where one starts a water journey, whether for only one day or for a month or more. At the other end of the journey is the takeout, where water creatures again take to land transportation and begin planning the next aquatic outing. Many of the put-ins that follow can also be the takeout if you make your excursion a round trip, ending where you started.

The Allagash Wilderness Waterway has much to offer, and it offers it in a variety of ways. Any of the put-in points up to and including Churchill Dam allow you to spend some time on the waterway and then return to your vehicle, thus eliminating the need for vehicle transport. This includes an Allagash Lake trip unless you wish to paddle from the lake back to Chamberlain Bridge. In that case you will have to arrange for your vehicle to be moved to the bridge. There are miles and miles of beautiful shoreline in the headwater lakes to be explored by canoe. Campsites that are a little out of the mainstream of down-river trippers are little used and offer a pleasant, quiet place to spend some time.

If your choice is a trip through all or part of the waterway, the following information about the various put-ins will give you some insight as to what is available and where to begin.

Many people have done the Allagash in stages, each year exploring a different part of the waterway, and more thoroughly than the folks who canoe the entire length in one visit. Read over the options as you plan your visit and choose the put-in that fits your needs and time available. Don't worry about what you may miss, because the Allagash will call you back again and again.

Telos Landing: The logical place to begin a chapter about starting places for an Allagash trip is at the beginning, or most southern point, of the Allagash Wilderness Waterway. Unfortunately, you can't get there! At least not by vehicle. There is no road access to the southern end of the waterway, so paddlers who wish to visit Telos must put-in at Chamberlain Bridge and canoe south.

Looking south from High Bank campsite on Telos. There are other views of the mountains from the waterway, but none compare with this one.

There are no state-maintained roads in the Allagash region, so anyone traveling to the waterway must go over private roads built and maintained for the purpose of transporting wood from the commercial forest that surrounds the park. Understandably, roads not needed for wood transport are allowed to go into disrepair, and the elements soon make them unusable.

In the 1970s, access to Telos was possible over some very rough roads, but soon the bridges and culverts were washed out and the roads were no longer available for vehicle traffic. But when they were open there were two possible approaches to Telos Landing. One was from the Telos Road where everyone travels today to approach Chamberlain Bridge. The other was through Baxter State Park. I had a very lucky experience using the Baxter approach.

After traveling through the park and approaching its border we came to a locked gate across the road. We got out and looked at the lock and could

Telos Dam. Here Allagash water is flowing south to Webster Lake and on into the East Branch of the Penobscot.

see that it was securely closed with a combination lock. We stood there, forlornly contemplating the many miles of backwoods driving that would be required to go around to the Telos Road and approach the landing from that direction. Just as we were about to turn around and begin the long, unanticipated journey we heard a vehicle coming from the other direction. We waited and a pickup truck stopped and a fellow jumped out and proceeded to unlock the gate. We yelled, "Don't close it until we get through," and jumped in our truck and headed out for Telos Landing. Sometimes lady luck smiles on the ill-informed.

If you have the time to do so, it is worthwhile to visit Telos if for no other reason than to make your exploration of the Allagash Wilderness Waterway complete. It will add one or two days to your journey. What appears to be all Telos Lake is actually made up of Round Pond and Telos, which run together as one. This is the product of the dams first built more than a century ago. There is a campsite on left shore of Round Pond called Chamberlain Bridge, a half mile or so from the bridge, where you could camp if you arrive late in the day. Farther down in Telos itself there is a campsite on the left side called High Bank. Dot and I agreed that it is the most picturesque campsite on the whole Allagash. There is an unrestricted view to the south of the lake and the mountains beyond, including Mount Katahdin, Maine's highest.

As you continue south on Telos the lake will narrow where two campsites lay opposite each other. On the right is the one called Telos Landing and opposite it is Field. Like many of the named places along the waterway, Telos

Landing is named for its former use, which was a place of access for lumbering operations. A walk through the woods around the campsite will turn up traces of the old lumber camps that were last used there in the 1930s.

A couple more miles down the lake is Telos Dam. Here was built the original dam, then known as the Dam-in-the-woods (see Chapter Nine, page 90) because it was originally built 165 yards from the water's edge in anticipation of the water rising with the completion of Chamberlain Dam at the site of present-day Lock Dam on Chamberlain Lake. Telos Dam is also a jumping-off point for a trip down the East Branch of the Penobscot, a canoe trip of considerable difficulty because of four major portages and other places that may cause difficulty depending upon the water level.

Telos Dam, along with its companion, Chamberlain Dam, succeeded in 1840 in turning southward into the Penobscot River, water that had flowed northward for thousands of years.

A lot of folks have trouble with the pronunciation of Telos (tea-loss). It is easier to remember if you look just to the west of the lake, where some wag named a little pond Coffeelos. In the early 1920s George S. Kephart* and a party of timber cruisers noticed a small unnamed pond in the vicinity. In a good-humored effort to complete the beverage theme, they dubbed the little pond Cocoalos. Unfortunately, future mapmakers failed to see the humor in this and the tiny pond became known as Little Coffeelos Pond.

Chamberlain Bridge: This is the most easily accessible and most often used put-in point for Allagash canoe trips. There is a ranger station near the bridge and a broad water access point just to the right of the bridge. There is also a boat landing here because large motorboats are allowed on Chamberlain and Telos lakes, unlike all of the park's other waters.

As you start your trip north from the bridge you will be in a thoroughfare for about a half mile, and just before entering Chamberlain there are two campsites. On the right is Boy Scout, and on the left is Thoroughfare. I have spent a lot of time in those campsites.

On one trip I had a group of eight people who were enthusiastic and eager to start their Allagash adventure. When we arrived at the bridge a northwest wind was blowing, and I knew right away that it did not bode well for the beginning of our trip. We struggled against the wind to Boy Scout and from there could see the huge waves rolling down the big lake. I informed everyone that we may as well set up camp because there would be no canoe travel on the lake that day. Even the ranger, in his big Boston Whaler, was reluctant to head up the lake and only did so once to make sure no canoeists were in trouble.

Normally, if you sit out a big blow like that one you can get up the next morning and have relatively calm paddling for a day or so. Not so this time.

*Kephart, George S., *Campfires Rekindled*, Channing Books, Marion, Mass., 1977.

The wind continued through the night and all the next day. There was nothing to be done. My guests were chafing to get going, but that was impossible. I constantly updated my itinerary in my mind, figuring how I would make progress once the wind finally went down.

Another night was spent at Boy Scout, and in the morning the wind was still rolling the waves down the lake. I was really starting to worry about getting the trip finished on time, and my guests were bored and irritable sitting around with nothing to do. Finally late that afternoon the wind died down and Chamberlain started to calm. I fed a quick meal and ordered the camp broken down. We headed up the lake late that afternoon and paddled until dark. It took two extra-long days of paddling after that, but we finally caught up to where we were supposed to be at that stage of the trip, and managed to end the trip at the appointed time.

Sometimes there are stories of how crowded the Allagash is. Weather like I just described is the root cause of those stories. If several parties arrive at Chamberlain Bridge when the wind is blowing, no one can move and the parties pile up. Since most people move along at about the same pace, people will likely be bunched up for a day or two until they naturally start to spread out. This sort of bunching can occur any time unfavorable weather causes people to sit tight rather than moving on, especially at a place like Chamberlain Bridge, where people can keep arriving by vehicle but cannot move on by canoe.

Because of that trip, and several others with similar stories, I decided that my guests would be a lot happier bypassing Chamberlain. I elected to start my trips on Eagle Lake by putting in on Indian Stream.

Indian Stream: You get to Indian Stream by crossing Chamberlain Bridge and continuing on the road for another five or six miles until you find a road on the left leading to Indian Pond. When you reach the pond you will see a campsite, but you are heading for the stream, so continue on until you reach a turnoff to the left, which is the road to Lock Dam. Access to the dam will be blocked just beyond the stream. There is a parking lot here for vehicles.

Indian Stream is not always a reliable access. I came in here early one fall with a party and found the stream nothing more than a rocky bed with hardly enough water to dampen the rocks. What to do? I had made arrangements for our vehicles to be picked up here and transported to Allagash Village, where we were to end our trip. It was obvious that we could not start our trip here, but I didn't know how to get word to my transporter about the change in plans.

I decided there was nothing to do but to go back to Chamberlain Bridge to start our trip and hope for the best. We drove back out past Indian Pond to the main road, where I wrote a note to the transporter and placed it on the signpost, hoping the drivers would notice it. When we got to the bridge I explained my problem to the ranger there. He said, "That's no problem, why don't you call them up?" A telephone had recently been installed at the ranger station, and a quick call solved my problem.

Like other roads in the working forest, the road to Indian Stream may be very good or very bad. It all depends on the work going on, or lack of it. Also be aware of new roads that may cause confusion as to which way is correct. Usually the correct one is the one showing the most travel, but not always.

Heading in to the stream early one summer, I suddenly came upon a fork in the road that had never been there before. That was confusing enough, but both choices showed new construction. The old road, whichever it was, looked just like the new one because of recent repairs. I stopped and told my guests to remain there—I would follow my best guess, to the left, and come back for them. As it turned out, my guess was correct, and I met someone heading out who agreed to tell my guests to follow, saving me the trip back.

The put-in at Indian Stream is just below three large culverts, and there is room to load, at most, two canoes at once. If the water flow is good you may be able to ride in the canoe along with your gear, but many times it is easier to wade along, controlling the canoe with the lines. It is less than half a mile to the shore of Eagle Lake, so this isn't a real hardship. Even at high water the trip down the stream is mostly a matter of bumping from one shore to the other because of the narrow channel and overhanging trees and bushes.

Throughout much of the history of lumbering in the Allagash region, well into the mid-1900s, the lakes and rivers were used to transport wood. As this wood floated in the water, sometimes for long periods of time, bark loosened and floated free. On many lakes in Maine the southeastern shores still have the accumulated bark from all those years of log driving, blown there by prevailing northwest winds. Bark is slow to decompose, and so you will see it on the bottom of Eagle Lake as you enter from Indian Stream. It appears to be black muck, but close examination will show it to be particles of bark.

The wind on Eagle Lake is not as serious as it is on Chamberlain Lake. This is because of the irregular shape of Eagle, which makes it possible to avoid the most open parts of the lake by following the lee shore as you make your way to your planned campsite. If the wind is blowing so hard you do not wish to go onto the lake, you could go back to Indian Pond to camp, or head to a closer North Maine Woods Campsite just beyond the final turnoff to Indian Stream. The nearest Allagash Wilderness Waterway campsites on the lake are on Pillsbury Island, about one and a half or two miles away by water.

Mud Pond Carry: This is the most ancient of all routes to Allagash waters. Its use goes back thousands of years. The trail between Umbazooksus Lake and Mud Pond is worn so deep that it is actually a stream during rainy weather. The name of the carry is appropriate, especially if you arrive during rainy weather. That was the case on my first time over the old portage. But first read about what happened just before we reached it. The following is the story I wrote about part of that trip for a Maine outdoor magazine.

LUCILLE

Among the many positive things about canoe tripping is the relationships you develop with the people you choose to go with. If it is with family or close friends the bonds are strengthened and all are better for it, and if your companions are newfound friends, what surer way to get to know them than to go canoe camping together?

Those of us who offer ourselves for hire, though, find the situation a bit different. We and our guests come together in the wilderness as almost total strangers, and getting along becomes a do or die kind of challenge. Fortunately, nearly all who desire to take a wilderness canoe trip are the type of people who are easy to get along with and a pleasure to know. This has been my experience at least, but there are exceptions. Let me tell you about Lucille.

It all started about eight or ten years ago. I received a call from a lady inquiring about a canoe trip. I said I had some time available and asked how large a group she had to go. "Oh, it's just me," she said. Well, I wasn't too eager to go off to the woods alone with just one woman, and I knew for sure Dot wouldn't be thrilled about the idea, but business is business and I quoted a price. The cost was too high; in fact, while I don't remember the amount now, I do *remember that her total budget amounted to about my daily rate at the time. She sounded very disappointed, and I felt kind of sorry for her. After talking for a while longer, I ended up promising that if anything came along I would contact her. I knew nothing would come along at that price and so hung up thinking that was that. I proceeded to forget about Lucille.*

In those days, part of my summer was spent running contract trips for a summer camp. On this particular year the group I had coming was made up almost entirely of kids I had taken out for several previous summers. In fact, they were now too old for the camp, but the director continued signing them because they wanted to canoe with me. As a result, I had to come up with a program that would last most of the scheduled four-week period, so as to keep them out of the camp area and the director's hair.

The kids were nearly all experienced canoeists, having done several traditional canoe trips with me in previous years. I decided to combine the Moose, the West Branch of the Penobscot, and the Allagash rivers to make a three-and-a-half-week canoe trip. I made arrangements for air resupply about halfway through the trip, and as the time approached I had everything in readiness and was looking forward to the challenge.

About a week before the youngsters were to arrive I got a call from the camp director. He said he was having trouble finding a counselor to go along as a second adult. He wondered if I had someone who would fill the job. I said I didn't. Later I happened to think of Lucille. I hesitated, then decided, why not? It was his decision, so I called him back and told him about her and our previous conversation. I said I didn't know her, in fact had only talked to her on the phone. As it turned out he called, interviewed, and hired her. Looking back

I always wonder what little voice (that I chose to ignore) had made me hesitate, but that is another subject altogether.

The day soon came. The kids and I had our reunion, and I met Lucille. One look at her and I could have assured Dot she would have had nothing at all to worry about if we had taken a trip alone together. Looks are skin-deep, though, and I optimistically told myself she was probably a great person and I would enjoy her company for the next few weeks. This just shows how wrong you can be when you don't listen to that little voice.

In fairness to Lucille, I will say she was at a disadvantage from the start. The campers and I had established a good relationship in previous years, and she was kind of an outsider to the group. A reasonable approach for her would have been to keep a low profile while the young people gradually accepted her. Not Lucille. Trouble started the second day out.

When canoe partners are assigned early in these trips I make it clear that they may be temporary because of a variety of reasons, such as not keeping up, not getting along, or whatever. I may change them for the good of the group. Who was the first to complain about a partner? You got it—Lucille. The boy with her was the weakest paddler of the group and had the worst disposition. I tried to explain to her as diplomatically as possible that I was aware of the boy's shortcomings and that *was the reason I had put him with her in the first place, so he would be with an adult. I made no changes.*

There is an old Native American legend (which I made up) that says females must be protected from canoe upsets by wearing a rawhide around their wrist. The girls in my groups were always delighted when the thong was tied on their wrist, and they faithfully wore them throughout the trip. (Some even wore them for a whole year). After presenting the protective rawhide to our girls on this trip, I offered one to Lucille so she could join the fun. She was indignant at the sexual discrimination (?), called me a male chauvinist (she left off the pig, but I know she was thinking it), and refused my offer. Needless to say, I avoided this type of subject thereafter, but her attitude did nothing to endear her to the group.

For the next few days things went along fine, I thought. No major complaints, and the group seemed to be coming together fine. Lucille hadn't shown much leadership or been much help to me, but I knew the kids well and really had no trouble with them. I realized that I didn't see much of her when we were in camp. I discovered the reason was that, except for meals, which she never missed, she spent all of her time in her tent. She couldn't stand the bugs.

One day an older camper came to me and told me what a pain Lucille was when I wasn't nearby. This didn't really make my day, because up to then I had counted Lucille as a zero. That is, not an asset to me, but not a liability either. But if she was making life miserable for the kids, then she was definitely on the red side of the ledger. She was nice as pie to me, but I started observing her around the kids, and she was definitely being a first-class pain.

She would get after them for little things that were insignificant and not what kids on vacation needed to hear from someone who, by now, they didn't even like. Any petty thing she could find to exercise what she thought to be her authority she would jump on. I wasn't sure how to handle it. Lucille was not the type who would understand a subtle suggestion that she tone it down a little. I dreaded a direct confrontation with more than two weeks of the trip still ahead of us. Still, if the kids are unhappy and don't have a good time, I have failed. To give myself time to decide what to do, I just kept myself between her and them, so to speak. Unexpectedly, the solution to my problem came from Lucille herself.

After all these years I still remember what she said: "I had a insatiable desire to paddle. I'm satiated." Not only that, but she was miserable with the bugs, the kids didn't like her, and besides she would miss her unemployment check if she didn't get out to sign up. She wondered if there was any way she could go out with the resupply plane when it came. Trying to hide my elation, I said that it would probably be possible to work it out and that I thought I could handle the kids alone, not adding that I had been doing that all along anyway.

When I told the kids about Lucille's plans they couldn't hide their joy as well as I did, and that didn't make relationships any better for the next couple of days, but resupply day finally arrived. The camp director himself came in with the plane to see how things were going, and needless to say he was not overjoyed with Lucille's proposal. However, when I had the opportunity to get him to one side and tell him how things were, he agreed to fly her out along with one canoe. I felt that a weight had been lifted from my shoulders.

There was celebration in Gil's camp that night! New goodies to eat, and Lucille was gone! Although we would finish the trip with a third person in one of the canoes, no one complained, and the kids agreed to take turns at it. This was quite a musical group, and there were the beginnings of a song about Lucille that night. It was complete a few days later. Unfortunately, I am not that musical and I don't remember the words after all these years, but you can bet they were sung with gusto around the campfire each night for the rest of the trip and for several years yet come.

All of the kids have graduated from college by now, but I hear from them now and then. None have forgotten Lucille. None have seen Lucille. None have missed Lucille. You did *pick a fine time to leave me, Lucille.*

You approach Mud Pond Carry from the south by traveling north on Chesuncook Lake, up Umbazooksus Stream, and onto Umbazooksus Lake. There is an old dam there that is no longer in use. The carry trail is not readily seen from the dam, but it is located about a mile away on the northwestern shore. Up the trail a ways is a campsite where most travelers set up in order to make the carry in stages. This is not an official campsite, as far as I know, so you should have a fire permit if you plan on staying there. It is probably little used.

This is Mud Brook, flowing from Mud Pond to Chamberlain Lake. The low water level here would hardly float a canoe, but you can see how the stream channel has been cleared by canoe travelers over the years.

You can read more about Umbazooksus Lake in Chapter Thirteen on page 144. It was the terminus of the Eagle Lake West Branch Railroad, and millions of cords of pulpwood were hauled here over rails laid in the late 1920s. Here, just as on the southeastern shore of Eagle Lake, the bark accumulation bears witness to the bustling activity that went on here so many years ago. A little poking around through the woods will also turn up artifacts of human activity that nature is trying hard to conceal and overcome.

The carry trail between Umbazooksus lake and Mud Pond is about two miles long and rises slowly from the Penobscot watershed to an indiscernible point where you will cross into the St. John (Allagash) watershed. Maybe it was just the weather during the times that I have been there, but I always had a gloomy, depressed feeling from the time I reached Umbazooksus Lake until I came out into the clear, shining waters of Chamberlain Lake. Maybe it was just all that mud. In Henry David Thoreau's *The Maine Woods* you can read about his adventures over this carry, when he and his companion were lost for the better part of a day. Pity the poor guide!

Mud Pond is well-named, and few desire to linger there. Mud Brook, leading to Chamberlain Lake, is almost directly across (north) from the end of the carry trail. At times beaver have dammed up the outlet, and if so you can open the dam and release additional water for your journey to the big lake. As you make your way down Mud Brook you will see evidence of the centuries of use. It does not look like most small brooks, but instead like a little canal. Over the years travelers have removed rocks that obstructed their progress from the stream bed and piled them on the shore. The little brook is best traversed by walking and wading along, controlling your loaded canoe by its lines. There is seldom water enough to allow people to paddle through. Besides, the stream is so small that it would be almost impossible to control the canoe by paddle or pole without constant bumping from shore to shore.

When you reach what appears to be Chamberlain Lake you can start paddling again, but this is really part of Mud Brook that has been flooded by what is now Lock Dam. A little less than a mile to the north you will see a narrows, and you will find a campsite on the left shore named Mud Brook. Ahead lies the largest of the Allagash lakes, stretching about 16 miles northwest to southeast.

Churchill Dam: People who wish to avoid the lake paddling can put in at Churchill Dam. From here to Allagash Village the trip can take from four to seven days, depending on individual preference, but five days seems to be about right. It is a long drive over dirt roads to get to the dam, so most people arrive sometime in the afternoon. There is a four-cell (four party) campsite on the west side of the river where you could spend the night and be off to an early start the next morning. The parking lot is also on the west side of the river.

If you have the time and wish to spend the night in a quieter place, you could paddle a couple of miles south of the dam and camp at Jaws or Highbank, then return to the dam in the morning. The ranger at the dam may be able to give you information as to whether or not those campsites might be full. Churchill is one of those places where people tend to pile up, because both folks doing the entire trip and people just putting in go through here.

In more than 30 years of guiding canoe trips I had to have only one person evacuated because of injury—it happened at Churchill Dam. She was running across the dirt road to get a photograph of someone starting down the river and somehow broke her ankle, probably by stepping on a stone. I guess if it had to happen this was a good place, because we had ice available to put on the break and a telephone to call for help. Because we were one person short in our party after her departure, I paddled the rest of that trip alone.

There is a ranger station on the east side of the river and the waterway headquarters is on that side as well, just before the dam. Safe drinking water is available from a well near the ranger cabin. You can get weather information from a ranger, and the forecast is posted daily on a small bulletin board near the dam on the west side. There you will find other waterway information as well.

Unfortunately for people without experience in white water, the hardest white water on the whole Allagash starts just downstream of the dam. I have always said the river was arranged backwards. If Chase Rapids were at the other end of the river few people would have trouble with it, because they would have had several days of easier white water in which to prepare. So, if you want to do just the river and avoid Chase Rapids, consider putting in at Umsaskis Bridge.

Umsaskis Bridge: The bridge is in the thoroughfare between Umsaskis Lake and Long Lake. Drive in past the ranger station to the put-in, and just beyond that is the parking lot. Nearby campsites are Sandy Point to the south and Pine about a mile to the north.

The trip from Umsaskis Bridge to Allagash Village takes three to five days. Four days is the usual. The only lake paddling will be Long Lake and Round Pond. All the rest of the water is Class I and easy Class II.

The road that crosses the bridge here is called the American Reality Road and can be accessed from the south, as with the other entry points so far, or from the east in the town of Ashland. In either direction you will pass through a North Maine Woods checkpoint where you will pay road fees and register and pay for your stay in the waterway.

I usually discouraged Allagash trips of this short duration simply because of the tremendous amount of driving, a lot of it over dirt roads, that is required for just three or four days of river canoeing. Not only is it a long way to the bridge to start the trip, but when you end up in Allagash Village, you are about as far from Maine's southern border as you can be and still be in the United States. Still, we did do it from time to time. It is a pleasant trip, and in my opinion Long Lake to Allagash Village is the prettiest part of the entire waterway.

Allagash Lake: This lake is a special place in more ways than one. It is maintained in as close to pristine condition as possible. The lake and the stream that feeds it are the headwaters of the Allagash. So, I suppose, those truly wanting to canoe the entire Allagash should start as far up Allagash Stream above the lake as possible. There is no vehicle access to the lake itself, and no motor transportation is allowed on it. If you hear a motor there it is the ranger doing her or his job.

Allagash Lake can be a starting place for an Allagash trip or a trip in and of itself. It also can be a side trip for the Allagash paddler who has time to make the six-mile climb up Allagash Stream to the lake and back again to continue north on the waterway. I will start with the upstream approach to the lake and move on to the easier ones a little later.

As you approach Allagash Stream, on the extreme northern end of Chamberlain Lake, the most prominent feature will be the old trestle of the

The trestle over Allagash Stream. An example of backwoods engineering that still stands today in spite of wind, ice, fire, and whatever else nature has thrown at it over the years.

Eagle Lake West Branch Railroad crossing in front of you. It has been there since the late 1920s, its endurance a tribute to the skills of those backwoods engineers and builders. Allagash Stream at this point appears to be a wide marsh because of the flooding caused by the dams. This man-made marsh is usually a great place to see wildlife, especially moose. Moose, like people, are individuals. Some are shy and will bolt for the woods at the first sight of an approaching canoe, while others will continue feeding, reminding one of cows in a pasture. Dot and I were starting a trip up the stream with our two daughters and some friends. Our daughter Julie was in my canoe and our daughter Kathy was in Dot's. As we paddled through the still waters we were surprised when the head of a large bull moose suddenly appeared beside us. The water where this bull was feeding on the bottom was so deep that when he put his head down for a mouthful of the aquatic foliage he completely disappeared. All that remained to mark the spot was the swarm of flies that patiently circled just above the surface waiting for their giant victim to re-emerge.

As you continue beyond the trestle, the distant woodlines slowly close in, and you will start to feel the current and find your progress slowed even as you pull harder and harder on your paddle. As you proceed up the stream the bottom will turn from sandy to rocky, and soon the current will be strong enough that further progress by paddle is impossible. This is the time to take up the pole or get out and start wading, pulling the canoe along as you do so. Poling requires a little experience and a lot of strength, especially upper body strength, and endurance. If you can pole upstream with a bow partner who is skilled at steering the bow of the canoe with a paddle, you can make progress much faster than if you must pole and steer both. As hard as poling

These are the two ways to "climb" Allagash Stream to Allagash Lake. Poling is faster, but requires some practice to master. Wading is slower, but anyone can do it if they have the patience and endurance.

is, it is much faster and easier than wading, in which you must push your body against the current while walking on slippery rocks and, at the same time, pull the loaded canoe with you.

Success at working your way up Allagash Stream requires an optimum water level. Too high and the task becomes nearly impossible, because the water is moving so fast that poling requires a huge amount of energy and wading is out of the question. Too low and poling is impossible in many places; you will be dragging the canoe over dry rocks in some of the shallower sections.

The preceding paragraphs do not give a very encouraging account of upstream travel, but I want everyone to know what they are facing when they decide to make the trek. On the positive side, though, this upstream work is very satisfying and will give you a small glimpse of how early explorers moved through the wilderness. For our Native Americans, going up and downstream was a way of life for uncounted centuries. You will read about this in Part Three, Native Americans.

Nearly in the middle of the six-mile upstream trip is Little Allagash Falls, where there is a campsite with the same name. Here you can stop and rest overnight in as beautiful a spot as there is on the entire waterway. You will fall asleep at night to the soothing roar of the falls. Camping here has the advantage of allowing you to make the upstream journey over two days instead of one. The portage around Little Allagash Falls is a short one.

Some friends of ours had an interesting experience at Little Allagash Falls. The couple was awakened early in the morning by a lot of commotion in their campsite. Crawling to the door of their tent, they sat and watched a full-grown black bear tearing, biting, swatting, and generally mauling their food pack trying to get at its contents. After some minutes of this (they said it seemed like hours), the bear gave it up as a lost cause and unceremoniously plunked the pack into their cold fireplace. In a foul mood at not finding an easy meal, his parting gesture was to clean off a shelf of the couple's cooking gear and utensils with one swipe of his huge paw. Without a backward glance, he lumbered away.

The food boxes were made from fiberglass* and were just too slick for the bear to get his teeth into. There were several tooth marks in the fiberglass surface, and one of the canvas handles had been torn off. The couple had the canvas handle replaced, but the tooth marks remain as a visible souvenir of their adventure. Bears are not a serious problem on any of the Allagash campsites because of the efforts of conscientious campers and diligent rangers to keep campsites clean and free of garbage. Still, it is prudent to take precautions to protect food by hanging containers out of reach. Also, be sure not to have snacks and goodies in personal packs that are in or near the tents.

* Gilpatrick, Gil, *Building Outdoor Gear*, Fox Chapel Publishing, 1999

When you are ready to continue your trip upstream, you will first cross Little Round Pond. This is one of the four Round Ponds in the Allagash Wilderness Waterway. The one near Chamberlain Bridge was already mentioned in the Telos section. The next one is between Eagle and Churchill Lakes, and the fourth one is farther downriver.

The trip upstream from Little Round Pond is pretty much the same as the previous day's climb. The pitch of the stream gets a little steeper as you approach the lake, but eventually you will get a glimpse of light ahead, signaling your approach to Allagash Lake. As you enter the lake there will be a campsite on the right called Outlet. Unless you are still filled with an abundance of energy at this point and wish to continue, you will probably want to camp here.

For many years the only way to get to Allagash Lake was to go upstream as just described or to carry overland for three miles from yet another Round Pond (not in the waterway) located south of Allagash Lake. This Round Pond is accessed by traveling from Caucomgomoc Lake up Ciss Stream. Then onto the north end of that Round Pond, where there is a carry trail leading three miles to Allagash Lake. I made the trip upstream several times, but never attempted the three-mile portage. Now there is an easier way to get to Allagash Lake, and it allows you to start your waterway trip here or to make this a trip by itself.

There are two possible ways to get to Allagash lake from upstream. The method in use for many years was by flying in to Johnson Pond and then working your way north on a small brook which empties into Allagash Stream above Allagash Lake. I have never traveled this route personally, and tales others have told of the difficulties encountered in traveling the small brook have never made me want to. However, until road access to Allagash Stream itself was available, this was the only "easy" way to get to Allagash Lake.

Look at a map and you will see where a woods road comes very close to Allagash Stream three or four miles above where the stream enters the lake. Since woods roads are in a constant state of change in the north woods, I will leave it to you to find the best route to the stream. People you encounter in the area (rangers, checkpoint attendants, etc.) can offer the best advice on the ground, and you will find that a copy of DeLorme's *The Maine Atlas and Gazetteer* is nearly indispensable. I'll digress here a moment to offer my method of navigating unfamiliar woods roads in the north woods.

There are few road signs on the dirt roads, and because of this it is hard to know exactly where you are on the map. New roads are constantly being built and old ones discontinued. To help keep track of your progress, sit down at home and choose your route on the map. Then, using the mileage scale on the map, determine the distance from a known point to a recognizable spot on the map along your chosen route. This spot might be a major stream crossing, a road junction, etc. Mark your measured mileage on the map at the recognizable spot, then continue measuring, making mileage checks every 10 or 15 miles until you reach a turn you must make, and then mark that. By using your vehicle's odometer you can keep track of progress and know when the turn-off you are looking for comes up. If you are careful your mileage will be within a mile or two of being correct, which is almost always close enough for you to be certain of your location. If you think you may travel that way again, you can change your measured mileage to the exact mileage at each checkpoint and turn-off. A vehicle compass is also helpful so you know you are heading in the general direction indicated on the map.

Allagash Stream above the lake is rather small, but large enough for easy canoeing as long as the water level is adequate. From time to time trees fall across the stream. These trees catch everything that drifts down the stream, creating a natural dam. These dams do not hold back much water at normal levels, but they do hold back canoes and so must be portaged. These obstacles are not predictable and may change from year to year. For many years there was a large dam in a bend of the stream. It had been there so long that a well-established carry trail had been created around it. Then the last time I was there it was gone, and we had clear canoeing all the way to the lake. There is no real fast water between the put-in and the lake, so it is relatively easy to travel back upstream to your vehicle if your trip is only to Allagash Lake.

As you enter Allagash Lake there is a choice of several campsites. The closest ones will be on your left. They are Ledge Point with three cells and Sandy Point with two cells. On the right you will see the ranger cabin, and a ways beyond that is Ice Cave campsite with two cells. There is little to choose among the campsites as far as a pleasant place to spend some time is concerned. I usually choose Ice Cave, if available, only because it makes a convenient spot from which to visit the caves.

From the Ice Cave campsite there is a well-established path leading to at least one of the caves. This cave leads downward at a fairly steep angle for 20 or 30 feet to a large chamber. Large is relative, but here it means there is room for several people to stand around and talk. I have never ventured beyond this point. On the way down to the chamber there is a place where you have to lie on your back and slide through a narrow opening. As I did this on one occasion I looked at the roof of the cave, about six inches above my face, and there was a bat hanging there. He was apparently used to these interruptions of his daily sleep time, because he was still there on my way out.

I always wondered about the name "Ice Caves." I never really believed that there was ice to be found there all summer long. In my first visit my skepticism was reinforced as I looked at the shiny, wet fungus that grows on the cave walls—it looks very much like ice in the glare of a flashlight beam, especially in the large chamber. For many years I was smug in the knowledge that I was correct about the source of the name. But then one day in late July my theory was shattered. I watched a young camper come down the trail with a chunk of ice the size of a large bucket. It could only have come from one the caves. So somewhere up there in that hill ice really does last well into the summer months, and an ice-skeptic had to admit he was wrong.

As you enjoy the peace and quiet along with the natural beauty of this pristine lake, I doubt you will become bored and need to be told of things to do. However, there is one man-made attraction that bears looking into: the fire tower on top of Allagash Mountain. From the tower you will get a magnificent view of the surrounding area and many of the lakes, including Chamberlain Lake (part of it). This high viewpoint gives one an appreciation of the distances involved in canoe travel in the area, as well as an idea of what an impact man has had here with lumbering operations.

Sukey explores the entrance to one of the ice caves. There are several of them, but this is the most accessible. You can quite easily get down twenty feet or so to where the cave opens up to a very dark chamber.

To reach the tower trail from Ice Caves you paddle about three miles down the western shore of the lake until you are almost all the way to the southern end. There will be a cabin on the west shore, and the trail up the mountain passes just to the right of it. The fire tower is not manned anymore, but it is open and you can climb up and have a look around. If you'd rather not go up into the tower, the view from its base is still worth the climb up the mountain. In August ripe blueberries can be found in places along the trail near the top.

Back in the 1970s, when Dot and I made this trip with our two daughters and some friends, the fire tower was manned every day throughout the fire season. At that time the lookout was Elmer Bowden from Dexter, Maine. The following is from an account I wrote about that trip.

"C'mon up," he shouted as he leaned out the window.

We did, and after crawling through the trap door in the floor of the little platform, we met Elmer. He answered a myriad of questions and had us sign his guest book.

The view from the tower was magnificent. We couldn't have asked for a better day. Nearly all of Allagash Lake was spread out before us, as on a map, and to the southeast, in a distant haze, loomed Mount Katahdin. We could see a small part of Chamberlain in the distance and were surprised to see how far we had traveled upstream to reach this wonderful place.

Elmer pointed out and named the other bodies of water that surrounded us like blue jewels dropped at random on a piece of green velvet. The closest of the little ponds were Upper and Lower Ellis Ponds. They were the headwaters of Ellis Brook, where many people camp on their journey up Chamberlain.

The trip back down the mountain was almost as difficult as the climb up, but not quite. A completely different set of leg muscles holds your back while going downhill, and it feels good when you finally get to walk on level ground again.

As we passed the cabin heading for our canoes, a lady stepped out and invited us in. She was Elmer's wife, Mary. We followed her into the cabin and sat down for a chat. The cabin was old, but the inside was brightly painted and spotlessly clean. Mary had been baking cookies and offered us some. In fact, she sat the pan in front of us. It was a treat for us, and I think Mary enjoyed our pleasure as much as we did her cookies.

You can leave Allagash Lake by traveling back upstream to your vehicle or by going downstream to Chamberlain. The outlet will be found almost due east of the inlet. The trip down the stream goes pretty fast, with only one or two portages. You will have to portage Little Allagash Falls. Then, below the falls about a mile and a half or two miles, there is a bridge where you should stop and have a look. You can decide whether you want to make the run under it or to portage around. The spot can be a little tricky, but if you and your partner are skilled in white-water canoeing you can probably make it with no difficulty.

As you approach Chamberlain the current will diminish, and soon you will be in the marshy area, with the old trestle visible in the distance. When you pass the trestle you will be in Chamberlain. On your left at the end of the lake is the terminus of the tramway, where some people carry across three-quarters of a mile to Eagle Lake rather than paddling down to Lock Dam and traveling that route into Eagle. If you are returning to Chamberlain Bridge, you will want to follow the west shore of Chamberlain, where most of the campsites are located along its 18-mile length.

[CHAPTER TWO]

CHAMBERLAIN LAKE

As you paddle out of the thoroughfare into Chamberlain lake you will see the Arm of Chamberlain stretching eastward for three miles. If you are looking for a quiet, secluded place to spend some time, there is a campsite about a mile or so up the Arm called "Arm." Few people use this campsite, as it is one of those out-of-the-way places.

If you plan to go through Lock Dam into Eagle Lake this day, then you may choose to travel up the east side of the lake, as Lock Dam is on that side. If you plan to spend the night on Chamberlain you should travel on the west shore because all of the campsites, up to Lock Dam and except for Arm, are on that side. About three or four miles up the lake, on the east shore, you will see Nugent's Chamberlain Lake Camps. These are commercial sporting camps, one of two grandfathered when the region became the Allagash Wilderness Waterway in the late 1960s.

Al (Nuge) and Patty Nugent first came to Chamberlain in the late 1930s, poling their way up the shore of the lake with all of their worldly goods piled on a raft. Together, with only hand tools and a lot of hard work, they built the cabins and other buildings necessary for a commercial sporting camp.

Farther up the same shore is Chamberlain Farm. If you plan to spend the night on Chamberlain and also visit the farm, you should plan to spend the night at either Ledge Point or Rocky Cove campsites, as they are located almost directly across the lake from it.

The most visible thing that identifies the farm is the large building that stands near the shore. As you steer your canoe for the wide beach near the building, your attention may be attracted to a large rusty metal object on your left in the middle of the south facing shore. That object is the remains of the steam side-wheeler H. W. Marsh (see Chapter Eleven, Page 117). The Marsh was the boat that had the stern cut off so the bow and midsection, where the engine was, could be pulled up on shore to keep the steam engine from freezing in. I guess, as it turned out, their efforts were a

The remains of the steam side-wheeler H.W. Marsh at Chamberlain Farm. Originally pulled ashore to keep the engine from freezing in.

waste of time because there it sits, a rusting monument to the wilderness navy of another century.

The big steam engine is a fascinating thing to look at and crawl over. The metal hubs of the two huge paddle wheels stick out to each side, their wooden paddles long since rotted away. Forward of the engine, in the bushes, are remains of some of the forward part of the hull. Most of the wood has rotted and burned away, but some parts that have not been in contact with the soil have remained in sound, though weather-beaten, condition.

Walking along the shore toward the only standing building at the farm, you will see, among the metal debris in the water, a long shaft and at the end a broken propeller that must have been at least three feet across when it was in use. This probably is what remains of the tow boat S.S. Dugan which was used to tow booms down to Telos Dam. The old photo on Page 94 shows a boat pulled up on shore that could have had a screw of this type; probably the S.S. Dugan.

The only standing building at the farm is now used as a camp by Nugent's. The forest has pretty well reclaimed the fields, but at the right time of year, the remaining ones are just bursting with beautiful plump wild strawberries. The only other building at the farm in recent years was an old barn, but that has collapsed and all that remains are weathered boards and timbers in a tangled mess.

Up the hill from the barn are foundations where other buildings stood. These are marked by square plots of birches and poplars, which usually indicates that the buildings were burned. One patch of the trees proved to be what was, in all probability, the main farmhouse. There was a deep cellar hole with carefully laid stones forming the foundation walls. On the uphill side there was a sloping ramp-way that formed an outside entrance to the cellar.

The only remaining building at Chamberlain Farm. It is used now as a camp by Nugents. There were, at one time, 600 acres cleared at the farm.

One has to wonder how many bushels of apples, potatoes, and root crops were wheeled down that ramp and stored in the cellar for winter use.

Thoreau stopped here at Chamberlain Farm on his 1857 trip, mainly to replenish his sugar supply. It seems that his Indian guide liked a little coffee in his sugar. Thoreau wrote, "He would first fill his dipper nearly a third full of sugar, and then add coffee to it." They obtained four pounds, all that could be spared, of brown sugar and were charged twenty cents a pound for it, "which certainly it was worth to get it up there," Thoreau added.

Lucius Hubbard* also stopped here at Chamberlain Farm in 1881. He described it as an extremely prosperous and productive farm that "raised just as good corn as grew anywhere in the state of Maine." About his encounter with the farm manager, he wrote, "So well do the sheep thrive there, that a short time before our arrival one became so fat that, in the words of the superintendent, Mr. Nutter, they had to kill him to save his life, couldn't lug himself around."

The point of land at Chamberlain Farm has yielded evidence of use of the area by indigenous people. Archeologists have turned up artifacts here in spite

*Hubbard, Lucius L., *Woods and Lakes of Maine*, Reprint New Hampshire Publishing Company, Somersworth, N.H., 1971.

of the fact that the present water level has most certainly covered much more as-yet-undiscovered evidence of Native American use of the area.

Back to the west shore now. After traveling up that shore for about four miles, the lake suddenly widens as Mud Cove comes into view on the left. My best friend, Dick Mosher, and I had quite an adventure in this area. This all happened back when I was still taking camp kids on trips, and on this one there were more people than one guide could legally have in a party. Dick agreed to take half of them, and we traveled more or less together, although we were equipped to camp as two separate parties.

As we started up the west shore of the lake there was a slight breeze blowing directly in our face, but it barely created a ripple on the surface of the lake. We were making good time, and the campers, their first day out, were doing as well at paddling as could be expected. Our destination was Donnelly Point campsite. When we were about halfway across Mud Cove the wind suddenly increased. I was in the lead, and the camper's canoes in my vicinity held the strongest paddlers. We pulled hard on the paddles and finally made shore at the campsite. About half of the campers were with me; some were from Dick's group, some from mine. The rest were blown toward the eastern shore, as they were unable make headway against the wind. Dick, seeing their difficulty, wisely stayed with them and ended up on that shore somewhere between Nugent's and Chamberlain Farm.

Extremely concerned with their welfare, I watched through binoculars as they were blown toward shore. My concern increased to near-panic when I watched one of the canoes upset. I could see one of my packs floating in the water, but everything else was so nearly submerged that I could see nothing of the people, the canoe, or other gear. Eventually I saw what I figured had to be Dick's canoe approach the swamped canoe, but it was too far away for me to see anything more than that. The only good news I could think of was that the campers had their own equipment on board with them, and I knew Dick had kitchen equipment. We each had food boxes, although they were mixed up, and the resulting meals would probably be a little different than originally planned.

I helplessly paced back and forth wanting to do something, but there were no possibilities other than to wait. The wind was now blowing too hard to attempt a crossing to check on the canoeists on the other shore. And even if I could have made it across, I couldn't go off and leave the campers who were with me, as the wind might prevent me from returning. Every few minutes I would climb on the highest rock on the shore and peer anxiously to the east with my binoculars. I could see people on the shore, but nothing that would give me any peace of mind. I had no idea if someone had drowned, was injured, or what.

Finally, late in the afternoon, Waterway Ranger Pat Reardon came up the lake making his rounds. I frantically waved my arms, and he came ashore. I explained what had happened, left a counselor in charge, and climbed aboard his boat. The big Boston Whaler quickly crossed the lake to the stranded campers on the other shore, and as it turned out they were all healthy and in good spirits. Dick had set up camp close to shore in preparation of spending the night. Dick and I checked out food supplies and then agreed to meet the following morning at Chamberlain Farm. Pat and I returned to my camp, my mind at ease for the first time in more than six hours.

Many canoeists like to spend the night at Ellis Brook campsite when traveling up the western shore. Ellis Brook is just a little south of being directly across from Lock Dam. The brook got its name from an old-time guide who was known simply as Old Ellis. Old Ellis had received ten dollars from a gentleman he had been guiding, with the request for a good bear skin. The following spring Old Ellis came upon the track of a bear, which he followed with determination for two or three days without catching up with it. However, he doggedly kept after it and was heard to exclaim, "Go it, old bruin, go it while ye kin! There ain't a hair on yer back that belongs to ye."

Lock dam today is a lock in name only. The dam is of earth and wood, and the water that makes it into Eagle Lake passes through a large pipe controlled by a gate located on shore. You will land on a gravel beach to the right of the dam, and it is just a short carry to the stream below. To the left of the dam is a cabin that has served as living quarters for dam tenders as well as Waterway Rangers. There is a four-cell campsite directly behind the landing area.

When Lock Dam was operational there were two dams that operated very much like canal locks. (See Page 96) With the upper dam closed and the lower one (Eagle Lake end) open, wood could be floated in between the dams from Eagle Lake. When the area between the dams was full of logs, the lower dam was closed. Next the upper dam was opened, letting in Chamberlain water and raising the logs to the surface level of Chamberlain lake, to be further transported down the lake, to pass through Telos Dam and into Penobscot waters and eventually to Bangor.

If you look closely to your right as you make your way down the small stream toward Eagle Lake, you can see old timbers in the water that are part of the old (lower) lock dam. A little beyond the old timbers you will reach Eagle lake level, and ahead will be Martin Cove. The distance you must travel to reach the lake level will vary with the water level of Eagle Lake.

On the trip I mentioned earlier where we intended to use Indian Stream, but couldn't, we found the trip from Chamberlain to Eagle Lake quite a challenge. Eagle Lake was extremely low that fall, and we found it necessary to portage our canoes and gear for nearly a mile. First it was down a carry trail beside the little stream (little or no water flowing) and then about a half mile across mud flats that are usually flooded over by Eagle Lake. That trip presented us with one challenge after another, most of them brought on by the low water level.

This large pipe is the outlet of Lock Dam. Usually enough water flows through here that it is an easy trip down the stream to Eagle Lake.

[CHAPTER THREE]

EAGLE LAKE

As you enter Martin Cove it is immediately apparent that the level of Eagle Lake has fluctuated from time to time. There are dead trees in the water that could only have grown originally on dry ground. In fact, the level of Eagle Lake has risen and fallen a number of times since the first dam was built about 1850. Like logging roads, dams were built and maintained as long as they were needed. When the need no longer existed, they were allowed to fall into disrepair. Then, as time passed and more timber matured, a dam was built again to assist in a new log drive. The present lake level has been maintained since 1968 when a dam was built, this time for the purpose of recreational canoeing.

Entering Eagle Lake from Chamberlain takes you into the southwest corner of Martin Cove. Coming in from Indian Stream will take you into the southeast corner of the cove. Looking north you will see a narrows that marks the entrance into Eagle Lake proper, and just beyond it is a large island known as Pillsbury Island. Pillsbury was the northernmost point Thoreau touched in his Allagash travels. He described and named the trees, shrubs, and other plant life of the island, concluding by saying, "I give these names because it was my farthest northern point." When Thoreau, his companion, and guide left the island they returned to Chamberlain, stopping at the farm, and then proceeding through Telos and down the East Branch of the Penobscot.

There are two campsites on Pillsbury Island. The one on the west side of the island is called Thoreau (four cells). The east facing site is Pillsbury Island (three cells). If you read Thoreau's writing about his visit to the island you will conclude, as I did, that the names of the island campsites have somehow gotten reversed. Thoreau wrote: "We landed on the southeast side of the island which was rather elevated, and densely wooded with a rocky shore." The southeast side of the island is where Pillsbury Island campsite is located. I think whoever named the campsites may have been confused by what Thoreau said later in the text: "In the meanwhile, the wind, increasing, blew down the Indian's birch and created such a sea that we found ourselves prisoners on the island, the nearest shore, which was the western, being perhaps a mile distant." At first glance it might seem as if he were referring to the nearest shore from their camp as the western shore. But what he *said* was that the nearest shore to the *island* was the western shore.

The campsites on Pillsbury are busy, as they lie in the direct path of travel of most paddlers. If you would like to go a little farther and stay at a somewhat more isolated site, paddle along the east shore of the island toward the mainland beyond. There you will find Smith Brook campsite (one cell). This is a very pleasant campsite under towering pine trees. It is also interesting that it has probably been a campsite for hundreds or even thousands of years. Lucius Hubbard, in his 1881 journey through the area, stopped at Smith Brook, and his writing suggests the spot may have been a favorite of the Native Americans: "Among other things, in plain sight from the lake, were three wooden frames made by Indians from the stems of young birch trees, and on which evidently had been stretched and dried the skins of some large animals." I have searched the shoreline near Smith Brook campsite and have found flakes of chert on the shore there where they worked in the manufacture of their stone tools. Who can say how long this place has been used as a campsite? I am sure if the water of the lake were down to prehistoric level many more artifacts would be discovered.

Unless you visited the tramway area from Chamberlain, that will undoubtedly be your next destination. To find the area, follow the southwestern shore until you find a narrow opening in the shoreline, which will appear like a narrow canal. The area is not readily visible from the lake and is easily passed. As you approach the area almost due south of Hog Island you are just about there. You can pull in the little canal and leave your canoe. From there the beaten path will indicate the way to go.

It is hard to imagine the tramway area the way it must have appeared over a hundred years ago. You can get some idea, though, by looking at some of the old photos in Chapters Eleven and Thirteen. Few people would apply the term "wilderness" to the south end of Eagle Lake back then. The area was cleared of trees and probably looked more like most any industrial area of the time. The thing to keep in mind as you visit the Tramway area is that it represents two different eras. The earliest, beginning in 1902 and from which it gets its name, was the tramway that transported long logs from Eagle Lake (north-flowing Allagash waters) to Chamberlain Lake (now south-flowing Penobscot waters, due to the dams). The later era, the Eagle Lake West Branch Railroad, started around 1928 and transported four-foot pulp wood from Eagle Lake to Umbazooksus Lake, also Penobscot waters but a different branch.

The first thing you see as you go ashore and climb a steep bank are the remains of some of the railroad cars used to haul pulpwood. Some wooden parts remain, but mostly there is just a jumble of metal parts stretched out through the trees and brush that has grown up in and around the rusting remains. Farther along are the two locomotives that hauled the wood, still sitting where they were abandoned in the early 1930s. Originally the old engines were in a shed, but it was burned and they have been exposed to the elements ever since.

The locomotives had been slowly sinking into the ground since being abandoned, and one of them was leaning alarmingly toward the lake. But a group of volunteers from the Allagash Alliance Group took on the project to put them right. They hauled in heavy jacks, blocking, and a huge variety of tools and equipment to do the job. Getting the tools and materials there was a major project in itself. Then they worked at getting the engines upright and stabilized. Working without the benefit of cranes or other heavy equipment, they slowly lifted the huge locomotives from the earth and put them on blocking awaiting tracks and a roadbed.

Next they dug, by hand, the soil from beneath the engines so a stable bed of crushed rock could be placed there, on which to place the ties and then the track. With that done, they now needed to get crushed rock to the site. The nearest place large trucks could deposit the material was over a winter road leading to a spot near Crows Nest campsite on Chamberlain Lake. The material was deposited there awaiting the next stage of the operation—hauling the crushed stone from Crows Nest to Tramway.

During the following winter the volunteers moved the equivalent of eight big dump truck loads (100 cubic yards) of stone from the Upper Crows

Terry Harper collection.

Volunteers from the Allagash Alliance Group starting the jacking of the largest of the two locomotives. It was tipping precariously toward Eagle Lake.

Terry Harper collection.

After the engine was up on blocks the group dug, by hand, underneath so the crushed rock could be filled in to provide a road bed.

Terry Harper collection.

Allagash Alliance Group member Tim Caverly hauls a load of crushed rock over the ice of Chamberlain to the tramway.

Nest area of Chamberlain Lake to the tramway. It was all done with plastic buckets, each weighing 80 to 100 pounds, hauled on snowmobile sleds and loaded and unloaded with human arms—arms the workers swear were six inches longer at the end of the two weekends of work. A total of 2,900 bucket loads were moved!

The following spring, with the stone on hand at the site, volunteers went to work constructing the under drain system and used the well-traveled stone to build the new roadbed beneath the locomotives. Again, the work was done by volunteers who traveled over many miles to reach the site and then put in long

The locomotives standing tall again after more than 80 years.

hard hours of labor while black flies and mosquitoes had their way with them. But the job wasn't done yet. There was not enough stone to build the road bed under one of the tenders, so the following winter the volunteers moved *another* 40 yards over the ice and snow. A labor of love for sure!

Just beyond the locomotives is a trail that leads three-quarters of a mile to Chamberlain Lake. Just to the right of the trail can be seen the remains of the tramway tracks, trucks, and cable. Half of the little two-wheel trucks are upside down. This, of course, is because they were at a lower level and were heading back for another load of logs when the whole thing was shut down for the last time in 1909. Nothing remains of the wooden structure that held the upper tracks on which the loaded trucks traveled. The cable and trucks lead in a straight line up over the height of land and on down to the Chamberlain shore, where you will find the machinery that powered the lengthy conveyor.

Two identical wood-burning boilers stand side by side on rock and masonry foundations. These are 12 or 14 feet long, and their firebox is large enough, if the grates were removed, for a person to stand in them. One time Dot and I, there on a photo trip, crawled into one of the fireboxes to escape a sudden downpour. We were dry, if a little cramped, squatting on the boiler's grates. On the ends opposite the firebox were the giant smokestacks, at least eighteen inches in diameter.

On top of the boilers is a broken six-inch pipe which was the steam line from the boilers to the big two-cylinder steam engine. The rugged but simple old engine looks remarkably intact. It too sits on a rock-masonry foundation and looks as if all it needs is a charge of steam to start its valves a-popping and

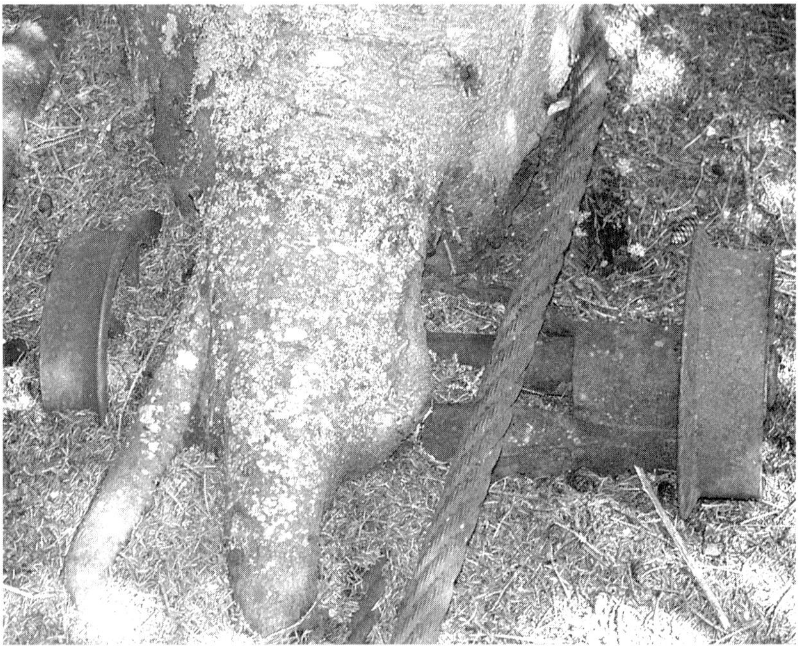

One of the tramway trucks. Nature has had more than 100 years to reclaim the tram machinery. It looks as if it will take at least another century.

its huge drive pulley whirling again. On Page 115 in Chapter Eleven there is mention of the crankshaft of the steam engine being broken, as it was being hauled in to the tramway area. If you look on the bottom of the little canal where you came in from Eagle Lake, you can see the old broken crankshaft in the water near the entrance to the canal. How it ended up on that end of the tram is anyone's guess. Perhaps someone used it as a anchor for one of the large boats.

The broken crankshaft from the old tramway steam engine. It is located in the water of the entrance canal to the tramway area, Eagle Lake end.

The tramway machinery located on the Chamberlain end. In the background are two huge wood-burning boilers that were connected by a pipe to the steam engine seen in the middle. The engine was connected to the foreground machinery by a wide belt that ran over the two wide pulleys.

This is the remainder of the drive machinery on the Chamberlain end of the tramway. The gears were turned by the large driven pulley in the previous photo. The cable with its trucks rode over the 9-foot pulley on the right.

The engine drive pulley is at least three feet in diameter and 18 to 24 inches wide. Apparently it had driven a tremendously large leather, or perhaps canvas, flat belt that had led from the engine to the drive machinery of the tram. The machinery is about 12 feet away, toward the Chamberlain shore. A small gear connected to the driven pulley meshed with a huge gear that served to mechanically reduce the revolutions per minute of the driven gear. Connected to the large gear is the nine-foot pulley, over which the one-and-a-half inch cable, the trucks, and the clamps rode to complete their delivery of logs to Chamberlain Lake, then turned back toward Eagle Lake in an endless rotation. The machinery and boilers all were inside a building when the tram was in operation, as can be seen in Photo on Page 114.

If you look to the left as you reach the height of land on your way back to Eagle, you will see a large boulder with an iron ring embedded in the top. I have never found out for sure the purpose of this ring, but I have a theory. There was a tower in the middle of the tramway where a person was stationed in a position to observe both ends. Probably there was a system of flag signals for communication. I think the iron ring anchored a guy-wire that helped to support the tower. I have searched the area for similar rocks but could find none. Probably they just took advantage of an existing boulder and anchored the other cables with "dead men." (A dead man is a log buried in the ground with a chain or cable attached to provide a firm, unmoving anchor point.)

When you leave the tramway area you will paddle straight up the lake, staying east of Hog and Farm Islands. Farm Island was a farm at one time, so that's how it got its name. Hog Island was used as a pasture for their pigs. The pigs must have needed constant care, because the small island could not have provided enough wild food to last any length of time. Farm Island still shows evidence of the subclimax trees, poplars and birches, that are growing where fields were cleared. You can see the same thing where there were clearings in the tram area. The light green leaves of the hardwoods contrast with the surrounding evergreens and mark, with ever-decreasing clarity, where openings once stood. The time for complete healing is very long in terms of human lifetimes. Hubbard had written about these islands in 1881, "On the sloping sides of one of these (islands), years ago, an incision was made and a budding farm ingrafted, but the wound has grown together again, and the scar is now scarcely visible."

Farm Island is a pleasant place to camp. I was camped there once with a group when we were treated to a beautiful display of heavenly fireworks. The night sky was ablaze with aurora borealis—the northern lights. I have watched many of these displays, especially during the fall and winter, but none that were a match for the one we saw that night. From where we were sitting around the fire we could see only overhead, because of the trees, so we went down to the shore of the lake where we could look to the northern horizon.

The lights flashed into the sky in streaks, starting at the horizon and shooting upward to nearly overhead. As one streak would fade another would flare up, and it was as if it were pulsing—a living thing. Directly overhead was the most unusual part. It looked like a starburst in a fireworks display, but the magnitude and beauty of what we were seeing made the most spectacular of fireworks seem very tame indeed. The "starburst" also pulsed and was aflame with color—pale pink and green. It was like a great umbrella of light directly over us. Between the "starburst" over our heads and the pulsing streaks from the horizon were thin clouds that looked as if they were lighted by the moon, but they weren't clouds. They were also aurora and seemed to jump from place to place in the sky. You could look at one for a few seconds and then it would fade. At the same time another would appear in a different place. The whole sky was alight with color and streaks against a backdrop of black star-studded sky. We watched the amazing display for so long our necks hurt. Then we stretched out on the rocky beach and just stared into the sky.

Beyond Farm Island the lake narrows and then opens up to what is commonly called Little Eagle—the northern lobe of the lake. On the northern shore of the peninsular that creates the narrows is the ranger cabin. There are several campsites on Eagle, most of them very good. If you want to get off by yourself, paddle to the northwest corner of the lake, where you will find Fred King campsite. Few people bother to paddle over to that out-of-the-way spot.

When the lake narrows again you will be leaving Eagle and entering a thoroughfare that leads to another Round Pond, then under John's Bridge and on into Churchill Lake.

[CHAPTER FOUR]

CHURCHILL LAKE AND CHASE RAPIDS

As you pass under John's Bridge you will be entering Churchill Lake. Actually, were it not for the flooding caused by Churchill Dam, you would still be in a thoroughfare for some time yet. Evidence of the raised water level is the dead trees in the water on your left. Also, you will see that the water is very shallow for a considerable distance out into the lake. Sometimes when the water is very low the area on the left is high and dry again. It was that way on one of my trips into the lake when we were treated to a rare sighting.

Nearly all of my guests on trips asked about the possibility of seeing a bear. Because the chances of seeing a bear on or near the water are remote, I never offered any encouragement in that regard. In fact, my standard reply was that in all of my years of running canoe trips I had seen only three bears from the water. I had been asked about bears by someone in this group only an hour or so before. As we approached the dri-ki* area, now high and dry, I could hardly believe my eyes. Out in the middle of the dead trees was a bear foraging for food. Then later, that same day, as we approached the other end of Churchill, we spotted a bear with a cub on the shore. So in that one day we saw as many bears as I had seen in all the previous years of guiding trips on northern Maine waters. Never say never or always, because Mother Nature will surely make a liar of you.

Nearly everyone who plans to continue downriver spends the night on Churchill Lake. The reason is that Churchill Dam is usually only opened in the morning. After it is closed the river is too low, all the way to Umsaskis, for canoeing. If you don't mind paddling for an hour or so in the morning before reaching the dam, the campsites called Scofield Cove (two cells) and Scofield Point (three cells) are pleasant ones. Farther on, closer to the dam, are High Bank on the left (one cell) and Jaws on the right (five cells). There is also a campsite at the dam called Churchill Dam (four cells).

**Dri-ki* is a Maine term for driftwood or dead trees in water.

Cow moose at Scofield Cove.

Wildlife is always around, whether you see it or not. It's a good idea to remember that the woods and waters are their home and we are the intruders.

However, that doesn't mean we cannot protect ourselves and our property when the occasion arises. We were camped at Scofield Cove; I had a large party. I was up making coffee and getting a fire started at the break of dawn. I heard a splashing down on the beach in front of our camp and knew it was a moose walking in the shallow water on shore. I hurried down to the shore to be sure that it didn't step in one of our canoes, which were pulled up on the beach. She was headed that way, and I ran out and waved my arms to shoo her away. She went to deeper water and bypassed the canoes, continuing on her way.

An old-fashoned wood splitter at Churchill Dam. There are other logging artifacts in the storage building.

Satisfied, I returned to my fire, impatiently waiting for my coffee to finish perking. A few minutes later the moose came back, crashing through the underbrush and straight into our camp. This time I didn't dare try to chase her off, because I was afraid she would spook and run into one of the tents, all of which held sleeping campers. I held my breath as she slowly picked her way through the tenting area, sniffing and looking around as if she owned the place. Which, of course, she did! Finally she disappeared into the woods again, this time for good. No one, besides me, was aware of how close they were to a 600-pound moose, although a couple of my guests did say they heard "someone" walking around their tent.

As you paddle from Churchill Lake toward the dam you will pass High Bank and Jaws campsites. When you do you will be entering Heron Lake. This is a small lake, only about a mile long, the north end being Churchill Dam and the beginning of Chase Rapids. This small lake is a favorite of moose. Several of them are often seen feeding on the western shore, especially in the morning.

The landing at Churchill Dam is on the left side. It is just a short carry from there across the road to a put-in below the dam. The state offers a portage service here that is inexpensive and well worth while for the insurance

The boarding house at Churchill Dam.

Present-day Churchill Dam.

it provides against wetting and loss of equipment. You can load your gear on the truck and then run the rapids with an empty canoe. Even seasoned white- water paddlers make use of this service, because even the best make mistakes—and besides, it is a lot more fun to run rapids with an empty canoe.

You will see a number of structures at Churchill Dam. On the left is the large boarding house that housed workers who worked in the area or who were in transit to or from one of the outlying lumber camps. On the right is a large barn-like building that could have served a variety of purposes. The other buildings are all modern ones for the people who care for the waterway. As you walk across the dam heading east, the headquarters building is in the

This was their first upset. They did it again a few minutes later, and one of them quit and headed back upriver on foot.

trees to your right. Across a field on the left is the ranger's cabin and beyond a maintenance building. In the field in front of the ranger's cabin is a well and a spigot where you can obtain safe drinking water.

Where you cross the road to the put-in there is a covered bulletin board with up-to-date weather information. This is always of interest to anyone who spends 24 hours a day outside. If the dam has been open for at least an hour, you are ready to head down into Chase Rapids.

As you approach the far end of the pool below the dam you can see a definite horizon ahead where the river narrows and drops sharply away. Stay to the left side of the pool, but head for the middle of the chute there. There may be a large rock exposed, or nearly so, in the chute. Keep just to the right of it. Below the chute the river widens somewhat but is quite steep, with large standing waves. If you go through the largest of these waves some water may splash into your canoe, but this can be quickly bailed out in the quiet water ahead. If you can, pull to the right side and avoid the big waves. Below is an area of relative quiet where you can stop on the right side and talk things over if you wish, and bail your canoe if necessary.

When you push off again, you can see the river is starting a gentle curve to the right. Head to the far left of the next drop, and as you start down it will became obvious that you should bring your canoe to the center of the river. There are some medium-size standing waves, but not many rocks to worry about. After a rather sharp drop and pretty good standing waves there are two or three hundred yards of good rapids with some rocks to be avoided, but no place that requires difficult maneuvers. A funny thing

Starting down the first drop of Chase Rapids. Churchill Dam is still visible at the far left.

happened in these rapids on one of my early spring trips. Funny to me, but not to the victims, I am sure.

We had met the group of young men a day or so before. They told me they did the trip every year. It turned out that some of them did it every year, but some of the members of their group were new and had not been here before. I guess I just assumed they knew what they were doing because they had been there before. I was wrong!

They started down Chase ahead of us. As my group started down the above stretch of white water, I looked ahead and saw one of their canoes upside down and the two men working to get it afloat again. I stopped and waited to see if assistance was needed. It was not, and they were soon under way again, though soaking wet and cold. I gave them some time to get well ahead and then shoved off again. When I reached the next set of rapids, the first thing I saw was a man walking up the right bank heading back up the river toward Churchill Dam. I yelled, "Are you OK?" He answered in the affirmative, although in kind of a grumpy way I thought. I continued down the river and the canoe in which the man had been in was not seen again until we reached Bissonett Bridge, where we got the rest of the story, and learned why the walker was kind of grumpy.

It seems the man we had seen walking was one of the new members who had apparently misplaced his trust in his veteran (?) partner. After the second upset in less than ten minutes he had had it! He wasn't going another inch! He left his group and walked back to the dam, I suppose. I never heard what became of him or how he managed to get home. Perhaps he later changed his mind and rejoined the group. I never found out, as we didn't run into them again.

Enjoying an exhilarating white-water run through Chase.

A flat area follows the rapids, and this is another good place to stop, talk, and bail if necessary. Just beyond is a shallow place where there were several channels, but they are small and do not align. You will go to the left through the shallow area. Look it over carefully and pick out the best channel. Once through the shallow area you will be on the left side of the river. Ahead the pitch will grow steeper, and there is a good white-water run for a hundred yards or so with fairly large standing waves. There are also some large boulders to be avoided, but there is plenty of room to maneuver, and few people have any difficulty here. At the end of this drop is another quiet area where you can stop again.

The river ahead disappears around a left bend. A map will show that this is the easternmost point in this section of the river. To start down the next pitch, paddle to the extreme right side of the river to clear a shallow area and then pull toward the center as you round the bend. Ahead are two or three hundred yards of good rapids, and the best course is slightly to the right of center. At the end of this pitch there is a grass island. The deepest water is to the right of the island, and beyond there is a pitch of two or three hundred yards of rapids that look steep and sound menacing, but will prove to be easier than the stretch just completed.

At the foot of this stretch of rapids is the Big Eddy, and it is truly an eddy. The current swirls both ways around the banks making it a natural catchall for articles floating down from the rapids above. Very often a short search in the overhanging alders around the banks of the eddy will turn up paddles and other gear that someone has lost in the rapids above.

A map will show that Big Eddy is just above the point where the water's course makes an abrupt right turn, starting the river on a broad loop to the east. The right turn is called Devil's Elbow. The rapid here does not seem significant enough to bear a name, but people frequently run their bows onto a nearly submerged rock, causing their canoes to do a neat 180-degree turn in the current.

The next one and a half miles is easy Class I and II, and you can paddle along leisurely, enjoying the large number of ducks with their families of little ducklings. These most likely are Mergansers, the most plentiful ducks on the waterway. The females (which are all you are likely to see in the summer) have a cinnamon brown head with an elongated crest. Their bodies are gray on top and whitish beneath. The male does not resemble the female. He has a black and white body with a green head and no crest. The reason the males are not observed in midsummer is that they are in their molt and cannot fly—they are in hiding.

Without the map it is difficult to tell you are on the large eastern loop, because slight turns in the river give the impression of a zigzag course. When you reach the easternmost point of the loop, the river swings to the left and there is a nice Class II drop of one hundred fifty or two hundred yards. There are some rocks to be avoided, but most everyone gets through without a bump.

The easy Class I river continues, now on a northwesterly course again, and in about one and a half miles you will pass a series of grass islands. In another mile beyond the islands the map shows a small loop to the east. When the channel straightens out and gets itself headed northwesterly again you can see the Bissonett Bridge abutment up ahead. There is no bridge there now, only the abutments. The ranger will have left your gear nearby. There is an easy set of rapids just above the bridge, and then you can land on the left side, both above and below the abutment where you will reload your canoe(s).

The river beyond the bridge is easy Class I all the way to Umsaskis Lake. As you near the lake, the first campsite you will pass is on the left opposite an island, a one-cell site called Meadows. In another mile or so there is a campsite on the right called Chisholm Brook (3 cells). Meadows is an isolated spot and will ensure you are alone. Chisholm Brook offers access to Umsaskis Marsh, which can be a fascinating place to spend an evening if you enjoy observing wildlife.

[CHAPTER FIVE]

UMSASKIS AND LONG LAKE

If you plan to spend a night on Umsaskis I recommend the Chisholm Brook campsite mainly for the opportunity to explore Umsaskis Marsh. It is also a pleasant campsite that very often offers up wildlife sightings without your even leaving camp. If you want to explore the marsh, you will have to paddle toward the lake for a half mile or so. Then you will see the waterway open up to your left. Just follow it into the marshy area. My daughter and I did this one evening, and the following is the account of our visit I wrote some time later.

We paddled along the river until we were nearly out to the lake. Then, off to the left, was a large opening and we paddled very slowly into the still waters. Ducks flew back and forth in small formations, and red-winged blackbirds perched on slender, swaying branches as we glided silently by. Their song, a gurgling konk-la-reee, o-ka-leeee was pleasant to hear, and that sound, along with others, was an important part of the total experience.

As the sun fell lower and lower toward the hills to the west, the entire area was bathed with a slanting, pinkish light that exaggerated every movement, every stirring of leaves and grass by the gentle breeze. We were caught up in it. We allowed ourselves to make no sounds, and for long periods the only things we heard of us was a faint swish of our paddles and an occasional gurgle as a bubble of air, pulled under by a paddle stroke, rose from under the canoe.

In the distance ahead of us a cow moose was feeding. We approached. When her head was down we would take several strokes with our paddles. When she raised her head to chew, breathe, and look around we would freeze and allow the canoe to drift toward her. When we were within a hundred yards she noticed us for the first time. We froze. She froze, ears forward, chewing stopped, a string of water lilies hanging from her dripping muzzle. For what must have been a minute the staring contest continued. Then, apparently deciding we were no threat, she continued chewing and dipped her head for another mouthful. Now each time she raised her head she looked directly toward us as we sat motionless except for our slow drift directly toward her. When we were about forty yards away she glanced shoreward. Following her

Umsaskis Marsh is home to a variety of wildlife.

look, we saw a calf standing in the brush. The light brown baby was far more nervous than its mother and moved out to join her. We didn't try to get closer. The cow took another mouthful. The calf fidgeted. Finally, after a long, almost disdainful look at us she slowly swung around and moved toward the nearby bushes. Her obvious nonchalance was comical. No way was she going to be hurried from her dining room! It had to look like her idea. The calf, in its relief to get out of there, almost entangled itself under her feet, but she would not be rushed. Three times she stopped to look at us before disappearing in the lush vegetation that grew along the water's edge.

As the sun dropped lower and the shadows grew longer everything became more and more beautiful. We paddled by a low mound of tall grass and something caught my eye. After looking a few moments, I saw the American bittern standing there, its beak pointed skyward and its long neck swaying back and forth in an almost perfect imitation of the surrounding grass. We later listened to its ka-choonk, ka-choonk, sometimes called the "stake driver," sounding every bit like someone driving a stake into the mud with a mallet.

We saw a great blue heron several times, but this shy bird kept a wary eye on us and never let us approach closer than a hundred yards.

Coming around a brushy corner in the quiet waterway I caught sight of a motion in the grass up ahead. There, looking at us over the tall grass, were the white-cheeked heads of at least a dozen Canada geese. We had heard their calls from our campsite and hoped to see some. As we continued, the flock disappeared, but around another bend they were swimming. They swam away from us, but as we drew nearer they finally spread their wings and left us to find a more desirable neighborhood.

The narrow area we had entered widened, and in a pond-like area a muskrat made a ripple wake in the quiet water while four ducks dabbled off to one side. They would up-end to feed on the bottom, then pop up, preen, and do it again. They were the very suspicious black ducks and at the first sight of us they leaped into the air with one powerful stroke of their wings and were soon out of sight.

The sunset this evening wasn't one of those spectacular colored ones we sometimes get, but the ever-changing light, both color and angle, gave the marsh a subtle beauty that is hard to describe and would not exist in the glare of the mid-day.

Our eyes and ears were not our only senses treated to the wild magic all around us. The variety of fragrances were nearly all unidentifiable, but I had smelled many of them before, and some triggered fleeting memories of when I was a boy. Even the marsh gas that bubbled up when we moved over shallow areas was not unpleasant as long as the doses were not too large.

When the sun dipped below the hills we retraced our route back to the main course of the river. As the water returns after being parted by a passing canoe, the marsh had closed in behind us. Everything seemed to have returned to the activity we may have interrupted as we passed. The cow moose was there, the geese were back swimming around, and the same noises persisted and some new ones joined in.

As we reached the main channel the frogs started up a steady cher-ump, cherump, and the canoe seemed to leap ahead as we picked up their rhythm and headed back toward camp against the slow river current.

From Chisholm Brook it is about a mile to Umsaskis Lake proper. The most direct route is across the lake toward what appears as a point of land about two miles away. If there happens to be a strong west or northwest wind blowing it is wise to take the longer route along the western shore. Because of the terrain surrounding Umsaskis, the lake can get very rough, in spite of its small size.

The point of land mentioned before is where you will find the Ledges campsite (two cells). Back in the 1970s I had quite an experience there.

I have experienced several hurricanes while on canoe trips, but the first one was the most memorable. Very often something good comes out of a situation that, at the time, seems terribly harsh and adverse. That happened to me on an Allagash trip with a small party.

The trip was booked by a lady for herself and her parents, who were probably in their mid-sixties. They all seemed in good health, and the trip was going well. They all had done some paddling before, so my job was easy in terms of getting the folks from one point to another.

When we reached Umsaskis Lake, about three days into the seven-day trip, we were notified by a waterway ranger that they were expecting a hurricane to strike sometime that night. I immediately thought of the high wind that we were likely to experience and so chose a campsite that would offer some protection with a minimum danger of falling trees.

My choice of campsite was good and the wind was not a problem, but the rain was the heaviest I have ever experienced while camping. We were set up on the side of a hill, chosen for good drainage because I knew rain would be coming. When I woke up in the morning the floor of my tent was rippling, virtually floating on a stream of water that was flowing beneath me. Most everything I had was wet, or at best damp. Moreover, I knew my guests would be in the same, or worse, shape. I crawled out of my tent in a downpour and hurried over to the tarp we had set up over the table and fireplace. I could see a fire was out of the question. I could have come up with dry wood, but it was raining so hard the whole side of the hill was one solid sheet of flowing water. Water was flowing right through the fireplace.

As soon as they heard me up and about, my party joined me under the tarp, where we discussed the situation. I learned then that Isabelle, the mother, had a respiratory problem that was being aggravated by the cold, damp conditions. They asked if I could find a way to get them out to civilization. As the weather showed no sign of improving, I didn't see any way to get a plane in, but I said I would try.

I went down to the shore and looked out across the lake. The wind had gone down some, and I decided I would paddle to the ranger's cabin to see if it were possible to get a plane in to pick her up. I didn't hold much hope for that possibility under the conditions, but I had to try. I went back to prepare their breakfast on my camp stove and make them as comfortable as I could before I went across the lake.

Just as we finished breakfast a very large ranger walked into our camp. We made introductions, and I learned his name was Brian Cain. He was actually an assistant ranger who lived in a cabin almost directly across the lake from our campsite. He was out checking to see how folks were doing after the big blow. I told him my problem and asked what he thought of the possibility of a plane getting through this mess.

Brian's answer was to invite all of us over to his camp to dry out and to stay as long as we wished. It didn't take me long to accept the invitation! One coincidence in this story is that Brian's camp was called the Gilpatrick Camp. It had been built by a man with my name, although I never learned whether or not we were related, or even what his first name was.

Brian had rounds to make, but described the camp location and told us to go over and move in, which we did. I soon had a hot fire going in the stove, and we spent the day drying out our clothing and gear. When Brian returned I prepared supper, complete with biscuits baked in the wood stove oven. We ate together and then shared the spirits that were available. My guests had brought along a supply of wine, and Brian had a bottle or two of something on hand as well.

What an enjoyable evening we had as we sat around that wood stove sipping our drinks and enjoying each other's company! We talked late into the night while the rain pounded on the roof and the wind howled through the trees around us. It was a good feeling to be in such a warm and cozy place in those conditions, because we knew firsthand what it was like to be out in them.

The next day the rain stopped, and we went out to see what an effect it had on the lake. I had dragged my canoes way up from the shore, but still the water was lapping at their sterns where I had left them. Brian's canoe was not quite as well stowed, and we had to take my canoe and go out on the lake to retrieve it. We stayed in the cabin another night to allow the water to go down a bit before we proceeded down the river. Even so, the water was so high and fast that we had no trouble making up for the unplanned time we had spent with Brian. The water in the river was running four or five times faster than normal. There were no rapids, just very fast moving water, because the high water had inundated all of the rocks.

That hurricane made it possible for me to meet what would turn out to be a lifelong friend and gave us all a wonderful memory. Neither Brian nor I ever forgot that time in the old Gilpatrick cabin. We always talk about it when we meet, which is usually every year or so. Brian went on to become a game warden and so left the waterway a few years after our hurricane meeting. He is a guy who could be truthfully called a "gentle giant," but I am not so sure how gentle he is with anyone who chooses to break a fish or game law. I do not intend to find that out personally.

Just beyond Ledges you will see the end of the lake where it narrows into the thoroughfare, which stretches for about two miles between Umsaskis Lake and Long Lake. As soon as you enter the thoroughfare you will see a bridge ahead and a ranger cabin nearby. This bridge is part of the famous American Realty Road, a private road that stretches all the way from the town of Ashland to the Canadian border at Daaquam.

Lining a canoe through Long Lake Dam. Very high or very low water can be a problem—better to portage. In this case, the water was a little low, so the canoe was helped over the timbers by the men in the water.

As you pass under the bridge you may be surprised to be propelled by a rather strong current, proving that, in spite of what appears to be still water, the lake is actually a wide part of the moving river. The thoroughfare widens somewhat just beyond the bridge and then narrows again, where you will again see and feel the current, before entering Long Lake.

At Long Lake I met another individual who became a lifelong friend. We were camped at Grey Brook, a two-cell site. In the middle of the afternoon, a gentleman came up the shore and asked if it would be all right if he and his group carried through our campsite to the empty one behind ours. Of course we welcomed them, and eventually the man came to us and introduced himself as Richard Barranger. We all shook hands, made small talk, and eventually they returned to their camp and we resumed our own camp activities. A little later Dot said to me, "His name is in the outhouse." "What do you mean?" I asked. She replied that his name was signed at the bottom of the waterway regulations that were posted there. He was the commissioner of conservation! Of course the waterway, a part of the Bureau of Parks and Recreation at the time, was part of his responsibility. We became friends, and he and his sons eventually built several strip canoes in my adult education program at the Skowhegan Regional Vocational Center. We usually kid him when we meet about his name being part of the reading material in every outhouse in the Allagash Wilderness Waterway.

Long Lake is little more than a quarter-mile wide at any point, but it is nearly four miles before you enter another thoroughfare, which is part of a marsh created where Chemquasabamticook Stream enters the waterway. Locally the stream is known as Se-bem-si-cook, or by its English name of Ross Stream. The narrow thoroughfare will eventually open up into Harvey Pond. There are campsites on both shores of Long lake and one on the western shore of Harvey Pond. Harvey Pond was named for John Harvey, who settled there in 1880. There is still evidence of his clearing about halfway down the pond on the western shore. Just beyond the old farm site is a campsite called Lost Popple, and beyond that the shoreline pinches in and the river falls away. Back to the river again!

In less than a mile of easy river paddling you will encounter Long Lake Dam. There is a campsite here by the same name (three cell). During rainy weather I often choose this campsite because it offers adequate shelter from the wind and excellent drainage, helping keep water from soaking tent floors. If you plan to camp here, or wish to portage, land on the right shore. If you are just passing through, it is often possible to line canoes over the old dam

This is the remains of a dump wagon that was used to build Long Lake Dam. It was loaded by men with shovels, then driven over the spot the fill was needed and the load released through the bottom of the wagon. Turn the page and see what was later done with this wagon.

timbers on the left shore. This makes it possible to continue your trip with a minimum of time and no lugging of gear. Look it over first, though—lining can be a dangerous practice in high water, and a near-impossible task when the water is very low.

Unless the water level is very high, it is not a good idea to run canoes over the dam. There used to be a channel where we did that, but then one year one of my canoes was ripped almost its entire length by a spike sticking up from one of the old timbers. Since then I have portaged or lined my canoes. The old dam is unpredictable, because each year the ice-out can rearrange things and move a hidden spike where none existed before.

There are a couple of river campsites beyond Long Lake Dam, Cunliff Island and Sweeney Brook (each one cell), each on the right-hand shore, a short distance below the dam. Beyond that there are none until you reach Round Pond.

Tim & Sue Caverly collection.

This is the same wagon I photographed years before. It was restored by Tom Goodyear of Ellsworth, Maine. It was a yearlong project involving more than 300 hours of donated labor. Tom also had to do some serious negotiating to come up with some of the missing parts. The restored wagon is located at Churchill Dam.

[CHAPTER SIX]

ROUND POND

From Long Lake Dam to Round Pond is a very pleasant section of river. It is easy Class I and II water with a few deadwater sections that seem like the river has stopped flowing, but take a look at the weeds on the bottom and you will see that the water is still giving you a good boost. For paddlers just learning to handle white water, this is a great section of river. There are places that require some maneuvering skills, but the mild current and shallow water depth usually ensure that a mistake does not result in an upset.

If you plan to camp at Round Pond, an early start is recommended. It is a popular spot with downriver paddlers who nearly all seem to plan to camp there. It is also road-accessible from Henderson Brook Bridge, so there may be folks there who drove in to camp for a few days. The early start will also increase your chances of seeing wildlife. Deer and moose like to feed in the river here, and they can usually be seen from some distance as you make your way down river. Again, remember that it is their home. Don't press in too close and drive them from their breakfast.

Osprey can often be spotted fishing along this section of river. Whenever I see an osprey I think of Leon. I first met Leon after receiving a call from the high school principal telling me something belonging to me had been left in his office. When I got there, I discovered it was my very expensive camera. I instantly realized I had left it hanging on a branch of a tree in the woods near where my class was building an outdoor classroom. Leon, who lived nearby, had found it and turned it in to the office.

I was so impressed with his honesty that I went to his home that afternoon and met his mother. Leon wasn't home, but I asked her if she thought he would like to take an Allagash trip with me so I could give some reward for the return of my camera. She said he would love it, and arrangements were made.

Leon became part of a group the following summer and ended up as my bow partner. Somewhere in the river between Long Lake Dam and Round Pond I looked up and saw an osprey hovering over the water not more than 20 yards in front of our canoe. We both stopped paddling to watch.

I had observed osprey hunting a number of times, but never this close, and I expected the bird to veer off because of us. It didn't happen, though! Suddenly the large bird folded its wings and dove into the water. It looked every bit like a jet fighter with swept-back wings. Or perhaps I should say the jet fighter looks like the diving osprey, because this bird's ancestors had surely been doing this long before aircraft designers came up with the concept.

The osprey disappeared in a splash and momentarily reappeared; then with one powerful stroke of its wings it cleared the water and was airborne again with a six- or seven-inch fish crosswise in its claws. The bird rose into the air in front of us to about 10 feet above the water's surface. Then its wing beats stopped for a moment and it shook itself violently, ridding its feathers of water in a cloud of spray. Next the large predator turned the fish it was carrying crosswise, so the fish faced to the front, thus making it aerodynamic. This done, the bird flew off to enjoy its catch or to feed its young waiting somewhere in a hidden nest.

Leon was so impressed with this display that he could talk about little else for the remainder of the trip. He was an eighth-grader when all this happened. During his last two years of high school he became one of my students and reminded me of that osprey experience from time to time. I was similarly impressed with what we had watched, and ever since that day, whenever I see an osprey fishing, I think of Leon.

There is a point in this section of the river where you may notice that you seem to have reversed course—that is, seem to be heading south instead of north. This is especially evident if the sun is shining. The river does take a turn to the southeast, and when it does you will start down through a long set of rips (a Maine term for easy rapids) that will end when you pass under the Henderson Brook Bridge.

Mothers have, through the years, offered sage advice that is handed down from generation to generation. Such things as "Never run with scissors," comes to mind. Another is, "Never stand up in a canoe." I will pause to take exception to the canoe thing.

I always stand up to pole or paddle when negotiating shallow water such as we encounter in the rips, especially during periods of low water. On one such trip I looked ahead and saw several individual canoes with the paddlers wading and pulling their canoes along searching for deeper water. With my knowledge of the river and by standing to seek out the deepest channels I soon passed these people with my group tucked in close behind me. When I approached Henderson Brook Bridge I looked back and discovered I had three more canoes in my group than I had started with. The inexperienced paddlers had temporarily joined my group to get through the shallow water with a minimum of dragging. They stopped and thanked me later, and I invited them to join us anytime—which they did from time to time farther downriver.

Standing is the most hassle-free way to negotiate shallow rips. It saves on the back muscles if your spare paddle is extra long.

Beyond the Henderson Brook Bridge the river starts to slow down as it approaches Round Pond level. You will see islands and several channels that you can choose. If you plan to camp at Back Channel campsite or Squirrel Pocket campsite, take the channel to the far right and it will bring you close to them. If you plan to continue down river or camp at any of the other sites, take the left-most channel. If you haven't made up your mind, take the center channel. Just before the river splits into the various channels, on the left is a perfect vase-shaped elm tree near the river bank. Thousands of elm trees have been killed in Maine, and the rest of New England, over the past 30 or 40 years by the Dutch elm disease. Somehow this one has managed to avoid infection. I look forward to seeing it each time I go through and am always relieved to see it is still healthy.

The silt from the rips above have built up the beautiful little delta where the river slows to enter Round Pond. There are a number of low islands that support a lush carpet of vegetation and created the several channels. Fiddlehead ferns are abundant, growing under the alders, a delicious fresh green when they first emerge in the spring.

Paddling down the right-hand channel one day, Dot and I were reminded that America's noble national bird, the bald eagle, usually classified a predator, is also a scavenger. Spotting an eagle is always a treat, and when this one flew up we quickly pointed it out to our guests. The wary eagle usually does not hang around when flushed like this, but this one flew into a nearby tree and sat there in full display. We wondered why this one was so bold until we saw (and smelled) a dead moose in the shallow water near shore. The eagle was not about to quickly abandon its easy meal.

Though it seems flat and calm there is a current through the islands. Moose and other wildlife are often sighted feeding here. You will enter the fourth, and final, Round Pond in the Allagash Wilderness Waterway. A few yards into the pond, you will be able to see nearly the entire shoreline.

Almost directly across the pond is a small peninsula projected into the pond, and on it are some camps. These are Jalbert (jal-bear) sporting camps, which existed before the waterway was established. Like Nugent's on Chamberlain Lake, they were allowed to continue operation under a grandfather clause. A half mile north of the camps is a small opening that used to be a Forest Service camp. The camp is gone now, and the opening serves as a campsite by the name of Tower Trail. It marks the beginning of a marked trail to the unmanned fire tower, which can be seen on a distant mountain. I would advise against camping at Tower Trail unless there is no alternative. It isn't a very good campsite.

As you enter Round Pond you will want to head for the left shore unless you are going to one of the campsites I mentioned above. All of the other campsites are located toward the outlet. In a half mile along that western shore you will see a tiny cabin that used to be a game warden's camp tucked into the woods. It is now used by waterway rangers. A short distance farther is a spring pouring into the lake through a pipe that came from a box built over the

Moose are often seen feeding around the islands where the river enters Round Pond. A visit there in the evening can be especially productive.

source. Like other springs along the waterway, this one has not been tested, so you drink at your own risk.

Many people who camp at Round Pond like to make the climb up to Round Pond Tower. It can be reached by hiking up a marked trail which starts at Tower Trail Campsite. It is about a two-and-a-half- to three-mile hike. The original tower was destroyed in 1978! It was similar to the burning of the shed over the locomotives at the Tram. It wasn't supposed to happen! The "who" and "why" of the event have been lost or covered up, and it is probably just as well. What is done is done.

Richard Barringer, commissioner of the Department of Conservation at the time, was furious about it. Forestry and Parks and Recreation were under his department. So Barringer himself, with his son Greg, some friends, and ranger Don Cote went up the mountain, disassembled the old tower and prepared the site for a new one. The new tower was brought in from another mountain. Within a year Round Pond visitors had a tower to climb again.

The trail from the east shore of Round Pond to the tower is not a hard climb. It goes up gradually, following contours for the most part, for about two miles. The last half mile or so is very steep.

You may not see as much of the river as you hoped from the top of the steel tower. You will see the islands where the river enters the pond and downriver you can see Musquacook Deadwater, but the rest of the water's course curves behind trees and is mostly hidden from view. The pond looks like a small puddle, and any canoes on it will appear as tiny specks. From time to time clearcuts can be seen, and that is sometimes a little depressing, though the forest heals fast and in a relatively short time the casual eye cannot even detect them. And remember, it *is* a working forest out there. Far to the south, you can see Mount Katahdin if the day is clear.

On the way down, after the very steep part of the trail, it is a pleasant walk through a variety of forest types. It is relatively cool and pleasant walking along the shady path. Spring-fed streams cross the trail in a couple of places.

We had camped early one day at Squirrel Pocket campsite. I had a lot of time on my hands in the afternoon, so I went out in my canoe alone just to pass part of the afternoon. I felt a pressing need to go ashore as I approached the Back Channel campsite. It was empty, so I went in and trudged up the hill to the outhouse. With no one around I left the door open to enjoy the view. Within a minute or so the view included a good-sized bear strolling leisurely down the hill about 5 yards in front of me. While I know bears are shy and avoid human contact whenever possible, it is still a little unsettling to be that close to one in such a compromising position. A little later, as I made my way back down the hill, the bear was there feeding on some berries in the empty campsite. He quickly disappeared as I approached. It is encounters like these that keep me going back into the woods. You just never know when the seemingly impossible is going to happen. When it does, you have a memory that will stay with you for the rest of your life—and a story to share with others to boot.

Another bear story before we move on downriver. I didn't see it, so it doesn't count for my total. Chris was on the trip with his brother. He desperately wanted to see a bear, but I offered no encouragement, telling him it was extremely unlikely. In the evening he and his brother went out for a paddle along the shore of the pond. When he came back, he was all excited about the bear they had spotted about two hundred yards from our camp. Of course I believed him, because I saw no reason why they would lie about it. However, Chris apparently was not convinced that I thought he was telling the truth, because about two weeks after the trip I received a large envelope in the mail. It was from Chris, and the 8-by-10 inch photo clearly showed a bear on the shore of the pond. He just wanted me to know, for sure, that he had indeed seen a bear on this trip. Chris has taken several trips with me since that time and he, his wife, Rose, and the rest of his family have become good friends.

When you leave Round Pond you immediately enter another set of rips. This one has been named Round Pond Rips, but is not any more difficult than the ones you encountered before reaching the pond.

[CHAPTER SEVEN]

ALLAGASH FALLS

Round Pond Rips starts immediately as you leave Round Pond and runs for nearly a mile. It is not a difficult run, easy Class II, and you can enjoy the descent watching alternately for rocks in the river and wildlife along its shore. The gradient of the river is more apparent here than in some other places so far, and it will make you feel as if you are sliding downhill, which of course you are!

There is always a chance for an upset in easy rips such as these, and inattention or inexperience is usually the cause. When it happened to a group I was leading many years ago, I *could* have said, "I told you so."

It was a small group of middle-aged men. As was my usual practice, back on the lakes several days earlier, I had urged them to get out on the lake with me to learn and practice the white-water canoe strokes they would need to negotiate Chase Rapids. Each time I brought up the subject they kind of mumbled and found something else that had to be done.

When the day came to run the rapids, I told them I was not going to let them do it. I said they were not ready and could upset and hurt themselves or damage the canoes. They were visibly upset with me. They were not pleased with my decision, but I did not give in, and they all rode around Chase in the portage truck.

A few days later when we were about halfway down through Round Pond Rips I heard a thump and an exclamation. I knew before turning around that someone had upset. It was not serious; I made my way back and helped them recover the gear and get the canoe upright, and we were soon on our way again.

That evening after supper, as we sat around the fire, one of the men said to me, "You were right back there at Chase; we were not ready." This simple admission by him made me feel a lot better about the whole situation, and it suddenly relieved the underlying tension that had been between us for the past few days.

Easy rapids continue down river for about two and a half miles. Then, where the river makes an abrupt turn east, it slows and deepens. This is the beginning of Musquacook Deadwater. The water is not dead here. It moves along with a good current, but it is a change over the fast-moving water above. The river flows almost due east for about two miles before hooking around to the north again. At about the midpoint of this easterly stretch there is a left bend, then a right.

The deadwater looks like a good place for moose, and it is, but there are never any guarantees. Sometimes you will see river bottom plants floating on the surface. When you see this, you can be fairly certain that moose had been there feeding a short time before.

At the point where the river turns north again Musquacook Stream enters from the east. This is where Lucius Hubbard reentered the Allagash on his 1881 trip. His party had left Churchill Lake and gone through the Musquacook chain of lakes, ending up on Musquacook Stream and following it back to the Allagash.

Musquacook is Native American for Birch Bark Place. There were probably birches somewhere along the stream that yielded bark for their canoes and utensils. It was not the way of Native Americans to give an entire river or stream a name. Rather, they named places along it according to some happening there. Probably some white man that mistook the name for a place as the name for the whole stream.

The river runs mostly slow and deep and almost due north for a couple of miles beyond where Musquacook enters the river. Then there is an easy rapid, followed by a campsite on the left bank called Hosea B. The name is taken from the name of a cabin that once stood there. The cabin was probably named for the brook that runs along side the site. Avoid camping at this site if you can. It is not a very good campsite, but makes a good place for a midmorning break in your paddling.

There is mostly deadwater from Hosea B to where Five Finger Brook enters the river from the east. There are three Five Finger Brook campsites: West, South, and North. Easy rapids will take you through the Five Finger area and around a small island near the northernmost of the campsites. Beyond, the river continues Class I and very easy Class II as it flows almost directly north. After about four miles you will pass campsites called Deadwater North and South, and after another mile, one called Bass Brook. All very pleasant river campsites.

You may find the rapids shallower than you have experienced before. It will become more and more important to locate the correct channel, lest you end up stuck on a shallow riffle with nothing to do but back up and find the channel or get out and walk the canoe through. A little farther down the river are two more campsites, one of which is also a historic site.

The campsite on the west shore is called Cunliff Bank, and one on the east shore is named Cunliff Depot. Cunliff Depot is the site of an old logging depot camp, and metal parts of old tools and machines litter the ground in many places. At one spot, near a small stream, the ground looks like and probably was an old junk yard. This must have been the site of a repair shed, and broken parts and other unwanted material were thrown out the back door. On the other side of the small stream, with trees and grass growing around and through them, are two old Lombard Log Haulers. One had a gas engine. The other, the older one, was powered by steam. The one with the gasoline engine had a cab at the rear and was steered and controlled from there. The other one, though, had the old steering wheel out front, where some poor soul sat and guided the machine and its train of heavy sleds as it slid and slewed over icy roads. (See Page 118.) The Lombard was the forerunner of the bulldozer and the army tank. The tracks on these old machines in the woods at Cunliff are not complete, but enough remains to get an idea of how they worked.

A couple of miles farther down river will be Michaud Farm. This was once a prosperous river farm, but now is the site of the final ranger cabin on the waterway. Here you should go ashore and check out with the ranger on duty.

Most people spend one more night on the waterway after passing Michaud Farm, although you may take-out there if you wish. The ranger at Michaud keeps track of where people plan to camp after checking out. By thus discussing the camping situation you can be reasonably sure that the campsite you are heading for will be open. It isn't a foolproof system, because some people change their minds and don't camp where they said they were going to. Still, you will have an idea of what is available and how many parties are ahead of you.

On a trip in the 1980s I was met at The Farm, as the locals call it, by Cecil Hafford, who was assistant ranger for the area at the time. Cecil was a large, gentle-looking, soft-spoken man who kept a small wad of Copenhagen tucked away in his cheek. I knew him from my previous trips. He said to me, without preamble, "You won't believe this, Gil, but we had someone go over the falls this summer!" He continued, "I have a picture of them going over, would you like to see it?"

Of course I wanted to see it, and with my group I followed him to the neat little log cabin, where he showed us the incredible photograph and told the story of what had happened. Cecil's story is not word for word, but it went like this.

"It was dark and I was at home when I got the call. Two people who were camped at the falls had witnessed the accident and had paddled and walked up to The Farm to get help. They said they didn't know how seriously hurt they were, but feared broken bones. I knew I couldn't leave

Remains of the Lombard log haulers at Cunliff Depot. Above: part of the gasoline-powered model. Below: an older one, steam-powered.

Photographer unknown.

A canoe with two people going over Allagash Falls. They were injured, but survived thanks to the high water at the time.

them there until morning, so I went up to the farm and took the canoe down to the falls in the dark.

"I was amazed that the pair were even alive and was eager to get them out of there and to a doctor. They didn't want me to take them out without their gear, though I told them over and over that the two of them and all the gear was too much of a load. I finally convinced them that medical attention for them was far more important than their equipment, which would almost surely be taken care of anyway.

"We set out, motoring back up the dark river, and when I came to where I cross over between two islands, right in the middle of the channel in front of us was a black bear! Those two New Yorkers had never seen a wild bear before, much less looked at one eyeball to eyeball, so they were pretty scared. I guess the bear didn't know what to make of all this commotion in the middle of the night. I didn't know if he was going to move or not for a while there, but finally he did. Those two fellers from New York had quite a day!

"I got them to the hospital at Fort Kent, and it turned out they only had some broken ribs plus some real good bruises. We picked up their gear the next morning.

"I don't know why they did it. I had warned them the same as I do everyone about the falls. If it hadn't been for that fellow from Rhode Island standing there taking pictures that afternoon when it happened, we would probably have trouble making people believe it."

A few years after that incident, there was another canoe that went over the falls. There were two people again—a man and a woman. The man survived with no injuries, but the woman was killed. Her body was recovered two weeks later about two miles down river from the falls. I never learned any details about this incident, except that it was initially deemed suspicious by authorities because the man survived without a scratch. Apparently it was eventually determined that there was no foul play, because I never heard any more about it.

The rapids at Michaud Farm are the last you will see until you reach the approach rapids to Allagash Falls. About a mile below the farm the river levels out and splits into a multitude of channels as it winds its way through a maze of islands. Approaching these islands is somewhat like approaching Round Pond. The lush vegetation, much different from that along the rest of the river, gives this portion of the waterway the appearance of a southern river. The predominant tree on the islands is silver maple, whose long branches hang pendulously over the quiet flowing river.

Finley Bogan is on the left opposite the first island. It was named for Finley McLennan, who had a farm here. Lucius Hubbard stopped at McLennan's seeking food and lodging after a long cold paddle down from Musquacook. McLennan told them "that he didn't keep a 'public house' and said they could get 'something' at Moir's, a mile below, where we could also put up for the night." The "something" referred to an alcoholic beverage that did not interest Hubbard. Moir's was the place called Moores Farm on modern maps; it is short distance below Finley's on the right-hand shore, where you can see the remains of the old log cabins.

The river flows around the alluvial islands for about a mile and a half and then, coming around a left bend, if the wind is right you can hear the deep, menacing roar of the falls. Soon afterward you can see the warning sign on the right bank informing you that you *must* portage.

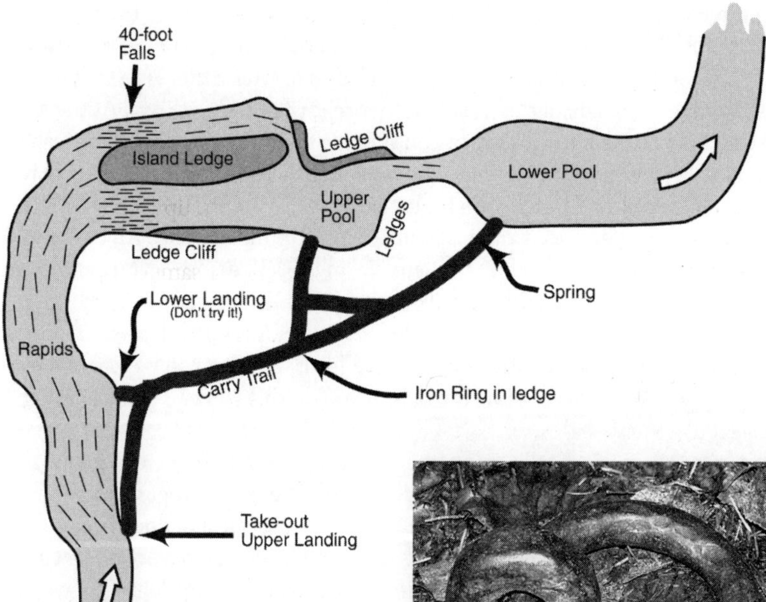

The rapids above the falls are not really difficult, and there is a place to take out a couple of hundred yards farther down, closer to the falls (lower landing). However, these rapids should command more respect than others like them because of the serious consequences if an upset should occur. The victims of such a spill would do well to save themselves from being swept over the falls. This means that the canoe and gear would be lost to the falls, or at best, recovered badly battered from the pool below. Even if items lodged on rocks before going over the brink, the recovery would have to be made from a very dangerous position. Most prudent paddlers carry from the upper landing, saving their white-water thrills for less hazardous locations.

Allagash Falls is another of those places that attracts people, and a large percentage of downriver travelers plan to spend a night there. So, if you wish to stay overnight, it is advisable to get there early to ensure yourself of a campsite. There are four cells here. One of the advantages of camping at the falls, besides it being a beautiful and interesting place, is that you can divide your portage into two days.

The carry trails around the falls are wide, made that way in the horseboat days (see Chapter Fourteen, Page 151) when wagons and even the boats themselves were rolled over them. The upper carry trail follows the river above the rapids. At the lower landing the trail turns right, and there is a boardwalk just beyond the turn. The boardwalk continues to where there is a ring in a rock on the right side of the trail--another relic of horseboat days. At the site of the ring you have the choice of continuing straight ahead or making a sharp left and down over a steep trail to the river below. The straight-ahead trail is a more gentle descent, but quite a bit longer.

There are several accessible viewing points around the falls. All of them have well-beaten trails leading to them. From the ledges above you can stand and look directly into the milk-white water as it roars and crashes over the jagged ledges to the pool forty feet below. It is difficult to see from above, but the falls split and form a large ledge island that can be reached by ferrying across the river below the falls. If you do this you can enjoy another spectacular view.

Children think the falls area is the greatest place yet. There are all kinds of "neat" places to climb, all of them designed to keep a mother's heart in her throat most of the time. Children have to climb, but that doesn't mean mothers stop the worrying. If you are so inclined, you can have a nice swim in the deep, fast-moving current fifty yards downriver from the falls. The water is swift and deep but it eddies around to a gravel shore below the ledges. You can swim in the current, and a couple of strokes toward shore will put you into the eddy. If you do this, wear your personal floatation device.

In the late evening I often sat alone looking at and listening to the powerful falls. I sometimes think about those two people going over it in a canoe. It was a miracle they survived. When I focus on the spot where the photographer caught them, I couldn't help but wonder what must have been going through their minds at that instant. Probably they don't even remember, but it must have been terrifying.

You are constantly reminded of where you are. The sound of the water falling over the ledges is everywhere in the area. Even when you walk a considerable distance to find dry wood, the roar is there like a guide to follow back to camp. It isn't just the sound, either. As you lay in your tent at night, getting sleepy to the soothing roar, you will probably became aware that you are feeling the falls as much as hearing them. Vibrations transmitted through the ledges can be felt for a considerable distance. Even a deaf person can "hear" Allagash Falls.

If you are interested in photography, you can get an early-morning shot of the falls as the rising sun shines directly on it. When you leave the falls area you will be entering the shallowest part of the entire trip, but in normal water levels you should have no trouble if you choose your channels carefully.

[CHAPTER EIGHT]

TO ALLAGASH VILLAGE

As soon as you round the bend below Allagash Falls you will notice a difference in the character of the river and countryside. The river runs wide and shallow for most of the fourteen miles to Allagash Village, where it enters the St. John, which itself becomes wider and shallower still. There are sections of relatively deep still water, but these only lull you into a false sense of security, and suddenly you find yourself at a dead end, with shallow riffles ahead and no choice but to backtrack or walk. Channels were blasted through these shallow areas in the log-driving days, and the experienced Allagash riverman follows them unerringly as he makes his way up or downriver. For the first-timer though, standing and paying constant attention is helpful in staying out of trouble.

About a mile and a half below the falls, at a bend in the river where there is a nice set of Class II rapids, McKeen Brook comes splashing into the river over the gravel bank. Just beyond is a campsite by the same name. It is a pleasant place to camp, with the sound of rapids and the brook to lull you to sleep at night. Most people I have guided through the years seem to name as their favorite campsites those with the sound of running water. The rapids just above McKeen Brook can be a challenge, and in low water it is important to carefully pick your channel, or you will come up short.

Just beyond the rapids the water quiets a bit and flows around an island, and then Big Brook enters from the east. Well-named, the brook cuts a wide swath through the woods off to the right. In another half mile you will encounter another island and the three Big Brook campsites. If you choose to camp at one of these sites you will have about a two-hour paddle to Allagash Village the next day. The time depends on the water level, of course, and with low water the time could be somewhat more. We had a very unusual bear sighting at one of the Big Brook campsites.

Whether a bear encounter is funny or not sometimes depends on whether you are the victim or an observer. We were the observers. It was mid-afternoon and my party was about ready to make camp at one of the sites north of Allagash Falls. From some distance upriver we could see one of the Big Brook campsites, and it looked like laundry day, with clothing and gear hanging everywhere. Since it had been sunny for the past few days, it was obvious that there had been an upset and they were in the process of drying out.

As we got closer we could see two young fellows sitting on the edge of the high bank with their elbows on their knees, their chins in their hands. A more dejected pair I could not imagine. As we approached, trying to think of something to say, we saw a bear come up behind them and start to rummage through their packs, which undoubtedly contained food. We yelled and told them what was happening, but they were having one of *those* days and without turning around said sarcastically, "Oh sure, that's all we need today."

After a shallow S turn below the campsites you will paddle past McGargle Rocks where the luckless log driver was killed while picking a log jam. (Page 138.) The ledges here offer little challenge to the canoeist, but were a bane to log drivers. The river continues alternating between easy rapids and easy current for the next four or five miles. You can easily pass Ghost Landing Bar without recognizing it, as there is no prominent identification. It is just before the last left bend in the river before a stretch of deadwater leading up to Twin Brook Rapids.

In the early days of logging along the Allagash, the only trees of interest were the huge pine trees that towered over the other trees in the forest. Two choppers worked with axes on one of the giants, and when it started to fall it twisted on the stump and one of the choppers was killed. The log was

Tim & Sue Caverly collection.

A large Allagash pine on the way to the river.

eventually hauled to the landing, to be turned into the river the following spring when the log drive started. But when the drive began, it was discovered that the log had a hollow core. It was abandoned at the landing. The story goes that the ghost of the chopper still sits on that log on the shore of the Allagash, beseeching passersby to turn the log in so as to set his soul free. You may see him there, but the log, like the earthly remains of the chopper, has turned to dust. You can do nothing but shrug and move on.

Twin Brook Rapids is the second-hardest white-water run on the waterway, though few paddlers have much trouble with them, having gained experience in the miles of river above. If you are not already wearing them, it is a good precaution to wear personal floatation devices when negotiating these rapids. There is a stretch of very deep water at the end of them, and anyone who upsets could get into trouble there.

I was running these rapids one time with my brother, David, in the bow. When we came abreast of the twin brooks I said, "Do you suppose it is two brooks flowing into the river or is it just one brook going across?" David cracked up! The comment struck him so funny that he was little help in controlling the canoe for the next minute or so. On the left bank, about halfway through the rapids, is a sign proclaiming the official end of the Allagash Wilderness Waterway.

A lot of folks wonder why this strange place was chosen to end the official waterway. It is because, at the time the waterway was being established as a state park, dams were under consideration for the St. John river that would have flooded the Allagash back to the location of the sign. If those dams had come to pass, you would be lake-paddling from that point to your take-out.

As you pass the last drop, where the river makes a right turn and there is a large ledge, you will see an island ahead. On the immediate left is a campsite accessible from a road that now parallels the river close by. Ahead, opposite the island on the east side, is a campsite called East Twin Brook. If you choose to stay here, you will be a little more than one hour away from Allagash Village the next morning.

This section of the Allagash is bordered by more hardwoods than you will have seen in the previous miles of your trip. Some of the hardwood growth is due to forest fires in the past 50 or 60 years. The forest has yet to reach its climax here. Some of it, though, is due to the natural selection of the hardwoods by soil and other factors. During a fall trip the foliage colors in this section of the river can be spectacular.

The six miles to Allagash Village will not be much different in character than the stretch down from Allagash Falls has been. I am always impressed here by the long stretches where the pitch of the river was so apparent. It really looks like a steep hill, though the current seems moderate.

The feeling of wilderness is often lost here because of the dusty road that follows the river's course. There is quite a lot of traffic up and down the road, some of it by trucks hauling logs, some by canoe outfitters with trailers piled high with canoes, and many private cars. Sometimes, if the wind is right, you may wish you had a dust mask to wear. After about two and a half miles you will pass an island opposite the North Maine Woods checkpoint on the private road that runs along the river.

If you have someone transporting your vehicles, that person will probably have established where the best place is for you to take out. As the land in the Village is privately owned, you should expect to pay a small fee for the privilege of taking out there. There may also be a daily fee for parking your vehicle.

Most people end their Allagash trip at Allagash Village, but it is possible to continue down the St. John River to St. Francis or even Fort Kent if the water level is adequate.

PART TWO

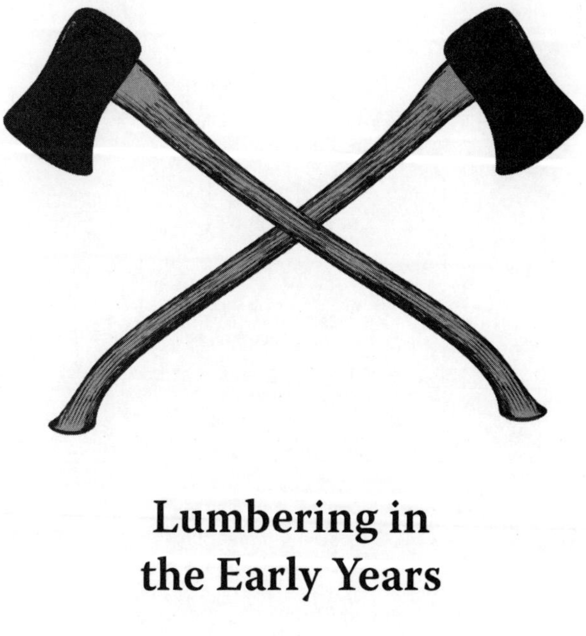

Lumbering in the Early Years

[PROLOGUE]

BANGOR

The importance of Maine's lumber goes back in history far beyond the time the first white man ever set foot in the Allagash region. In the year 1666 Samuel Pepys, Under-Secretary of the British Admiralty, made a report in which he foretold gloom and destruction for the nation. However, his pessimistic account closed with this item.

"There is also the very good news come of four New England ships come home safe to Falmouth with masts for the King; which is a blessing mighty unexpected, and without which, if for nothing else, we must have failed next year. But God be praised for this good fortune and send us a continuance of his favor in other things."

Four shiploads of Maine pine saved the mightiest nation on earth from destruction? Not likely, but Pepys' report does underline how vital this new-world lumber was to the mother country.

Allagash timber did not become important to anyone until the city of Bangor, some two hundred miles to the south, started running out of raw material in the Penobscot watershed to feed its hundreds of mills. By 1860 there were 410 lumber, shingle, and clapboard mills in the Bangor area. Waste alone was a tremendous problem. Enormous quantities of sawdust, slabs, and edgings went into the river and, at one time, blocked shipping to such an extent that the United States Government had to dredge it out. Individual owners of the mills entered into long and bitter lawsuits over the waste from one mill blocking the wheels of another.

Ships from all over the world arrived and departed from this river port, and they were often anchored so close to one another that a person could walk on them from Bangor to Brewer. Shipping to the West Indies was especially heavy. There, in return for the lumber, the ships brought back rum and molasses, a mutually satisfying exchange.

The lumber capital of the world was not all mills and ships. Its streets were well-equipped to serve the sailors coming upriver as well as the woods and river men coming down. The plank sidewalks were lined with saloons and houses of ill-repute, both of which did a brisk business.

"Why do you call that squirrel whiskey?" asked a logger of the barmaid who had just introduced him to the awful brew.

"Because," she replied sweetly, "it will make a lumberjack run up one side of a tree and down the other just like a squirrel."

The houses or ill-repute that lined Exchange Street advertised their services by displaying a sign reading, "Gentlemen's Laundry Done." Now and then a greenhorn showed up with an armful of dirty shirts and underwear, providing great amusement for the working ladies in residence.

Seeing their source of pine drying up caused a mild panic among Bangor lumberers, and Allagash land speculation ran rampant in the 1830s. Judging from a report in a publication of the time, someone could see some humor in the situation.

"It is rumored that one evening last week, two paupers escaped from the Bangor Almshouse, and they were caught early the next morning, yet in the meantime, before they were secured, they had made $1,800 each by speculating in timber lands. "

Powerful Bangor lumbermen like Sam Veazie and Rufus Dwinell were not the kind of men to sit idly by and watch their empires dwindle because of a lack of raw material. Nor were they the type to sit back and wait for someone else or the government to do something. They made things happen! They acted first and worried about the consequences later. Their mills could not be moved to the timber. The timber would have to be moved to their mills.

This then, is how a region, far away from civilization, came to feel the heavy hand of man.

[CHAPTER NINE]

DAM IT ALL

Maine had not yet been a state for twenty years when the Bangor lumber barons began to scheme ways to obtain Allagash pine to feed their greedy mills along the Penobscot. The first steps taken to turn Allagash pine southward began when the headwaters of the East Branch of the Penobscot were being cut over.

The Strickland Brothers and Amos Roberts owned Township 6, Range 11 (half of Webster Lake and all of Telos) and in 1839 were in the process of cutting pine around Webster Lake to be driven down Webster Brook and into the East Branch of the Penobscot.

One has to understand that these men were doers, not dreamers. The fact that there would be insufficient water from little Webster Lake to make a successful drive down the brook did not deter them from going ahead with their cutting operation. They would worry about the water problem when the time came.

Nor were the partners in awe of state government or concerned about legalities that might be involved in gaining the water they needed. The idea of turning Allagash water into the Penobscot was not a new one, and had been the subject of many heated debates in recent years among Bangor men. In fact, the state legislature had been petitioned to charter a company to build a canal to divert the Allagash water, but this was turned down and the matter laid to rest. At least that's what the state legislators thought.

Roberts and the Stricklands knew of a man named Shepard Boody who, besides being a lumberman, was an extremely able practical engineer. They hired Boody and sent him to investigate the possibility of obtaining Chamberlain water for their drive. This was the kind of challenge that Boody liked, and he readily accepted the job.

After a careful survey of the Telos-Chamberlain area, Boody found the only place from which Chamberlain water could be obtained for the Webster drive was from the head of Telos. Here he found that the land rose five or six feet over 30 rods and then fell away, until at 60 rods from the Telos shore it was at the low-water level of the lake. At this point there was a run or gully that dropped, over about a mile, 47 feet to Webster Lake.

Boody figured that a dam with an 11-foot head at the outlet of Chamberlain would provide sufficient water at Telos to overflow into Webster Lake without a cut. The only other necessary construction would be a dam at the head of Telos to control the flow.

It is extremely doubtful that these crusty lumbermen working in the wilderness of northern Maine needed anything to salve their conscience over turning water from one watershed to another. Nevertheless supporters of the project argued that God had originally intended Chamberlain water to flow through the run or gully he had provided into Webster Lake. An accident of nature had prevented it. They were merely carrying out what God had originally planned.

In the early spring of 1840 the partners, acting on Boody's recommendations, built what was then known as the "dam-in-the-woods." This dam was 30 rods back from the shore at the point where the land began to slope toward Webster Lake. At the same time a dam was built at the outlet of Chamberlain. With this dam to provide the needed 11-foot head and the dam-in-the-woods to control the flow, they expected to have enough water to make a drive.

Roberts and the Stricklands were certainly not considered fools. Yet the simple laborers who toiled on the dam almost 500 feet into the woods without a drop of water in sight must have been puzzled. Their comments, if they had been recorded, would make colorful reading. However, if Roberts and the Stricklands' esteem had gone down in the eyes of their employees, it was quickly restored in the spring when the ice went out and Telos overflowed through the dam-in-the-woods. Water from the Allagash was for the first time—at least for the first time known to mankind—flowing south into the Penobscot waters. The partners made a successful drive down Webster Brook, and their logs eventually reached Bangor. Not that anyone ever doubted they would.

Now the gates were open! Both figuratively and literally. Nothing, come hell or low water, would stop Allagash pine from reaching the hungry saws in Bangor.

In 1841 trees were cut above and below the dam-in-the-woods, and it became known as Telos Dam. A six-foot ditch was dug through the 30-rod bank above the dam, and obstructions were cleared from the ravine below to Webster Lake, creating Telos cut. In 1842 the first pine logs from the Allagash watershed were put through the cut and driven to Bangor.

The Allagash pine was cut by men on a contract basis for Roberts and Strickland Brothers. These operators paid the partners fifty cents per 1,000 board feet, toll for the use of the cut and the dam.

The partners operated the cut about two years. About $6,000 in tolls was collected. The cut became inoperative when Chamberlain Dam went out. The dam was rebuilt, but not high enough for water to flow through Telos Dam, and no logs passed from Allagash to Penobscot waters for the next few years.

The circumstances that caused the entire Allagash to truly feel the hand of man began in Salem, Massachusetts, where a wealthy young shipping magnate named David Pingree became, somewhat inadvertently, the owner of huge tracts of Allagash timber.

A wealthy friend by the name of Fiske stopped by Pingree's office one day, and in the course of their conversation Fiske unloaded some of his troubles on his friend. It seems that Fiske's sons were taking the rosy path of wine, women, and song. Fiske, searching out a way to divert their dalliance, purchased several townships of timber in some out-of-the-way place in Maine called Allagash. But, to Mr. Fiske's distress, the boys would have no part of it. Their idea of something to do definitely was not canoeing around the wilderness with a bunch of savages. Mr. Fiske was at a loss as to what to do with the boys and the timberland.

Pingree was looking for a place to reinvest some of his capital, but the idea of buying timberland in the wilds of Maine never occurred to him. The whole matter would have been dropped immediately except for the sharp ears of Pingree's office clerk.

The clerk had overheard their conversation, and he persuaded his employer to loan him the money to buy the land. The young man went directly to Maine and was extremely successful with his timber dealings. However, he did not confine his dealings to timber and soon found himself bankrupt. Back in Salem, Mr. Pingree found himself with several million acres of Mr. Fiske's timberland.

Not accustomed to taking a loss in his business dealings, and with no intention of doing so now, Pingree searched for a way to make the best of the situation. He remembered a man in town by the name of Ebenezer Coe. He was an engineer and a genius. Furthermore, if Pingree's appraisal of him was correct, he was energetic, shrewd, and would be extremely loyal.

It took only a short time to prove the wisdom of Pingree's choice of an agent. Eben Coe was a man who did things right. He hired canoes and guides and explored not only Pingree's land, but the rest of the Allagash headwaters as well. He was impressed with what he saw, and his guides told him that the stands of timber extended north down the river all the way to the St. John. His practical engineer's mind was pleased by the fact that most of the best pine stood near the waterways.

When Coe reported back to Pingree in Salem he was boiling over with enthusiasm. He advised Pingree not only to hold on to the Maine land he already owned, but to buy more. Pingree followed his advice and made Coe a full partner in the north woods enterprise. The engineer was to invest his time, brains, and energy. Pingree would invest his money.

To get the Allagash pine into the Penobscot, Coe intended to dig a canal between Mud Pond and Umbazooksus Lake. Then, by raising the level of Chamberlain with a dam at its outlet, logs could be driven into the West Branch of the Penobscot. This, Coe thought, was preferable to driving the East Branch, which was extremely kinky and prone to frequent log jams.

However, when Roberts and the Stricklands built Telos Dam, Coe realized the West Branch option was no longer open to him. He had waited too long. Now, although he and Pingree owned several Allagash townships and Chamberlain Dam, they were worthless without Telos Dam.

Roberts offered Pingree the key township containing Telos Dam, but he refused, preferring to have someone else operate the cut and feeling sure a reasonable rate could be established for driving logs through. Besides, it was obvious that Telos Dam was as worthless without Chamberlain Dam as Chamberlain Dam was without Telos Dam.

Probably all would have gone well if Rufus Dwinel had not appeared on the scene. Dwinel was a prominent landholder. Among Bangor lumbermen, if he was second to anyone, it was to General Sam Veazie, who had acquired enough property for a principality, along with 62 saw mills, the Penobscot Boom Company, and his own town just north of Bangor.

Once firmly established in his little monopoly, old Rufus became a staunch defender of individual property rights and insisted that to trespass on his canal was every bit as offensive as to trespass across his living room.

Eben Coe had rebuilt the dam at Chamberlain, expecting to make a reasonable drive through Telos canal. But this was not to be, and the first battle of what was to become known as the Telos War began.

Dwinel set his toll at 34 cents per 1000 board feet. When Coe-Pingree and others cried "robbery," he promptly raised it to 50 cents. He reasoned that their refusal to pay constituted a threat to drive their logs through by force. The additional 16 cents was to pay for the force of men he hired to defend his property.

Pingree and Coe fully realized the extent of their mistake in not acquiring Township 6, Range 11. Though they controlled Chamberlain Dam, it gave them little leverage or bargaining power. To allow Chamberlain water to escape through its natural outlet would be cutting off their nose to spite their face.

Rufus was not a man who made idle threats. He traveled downriver and hired 75 Bangor Tigers from Exchange Street in Bangor and brought them to his dam armed with knives, picks, handpikes, and axes. These fellows were more accustomed to brawling with fists, teeth, and spiked boots, but they were willing and eager to wield their new weapons in defense of their employer's property.

This first battle of the Telos War reached the critical stage when the Coe-Pingree operators, who were moving logs toward the dam, reached the thoroughfare into Telos. They fully intended to run their logs through Dwinel's dam, perhaps to test its legality, if not the strength of its defenders. They figured that any damages that might be awarded to Rufus would be considerably less than his tolls.

Dwinel was determined to stop them. His army of Bangor Tigers met the drivers as they approached the cut. Henry Cotton was one of the Coe-Pengree operators who were moving logs toward the dam. When he met Dwinel's chief deputy, Henry Head, Cotton offered to pay the 34 cents toll. The deputy informed him the price was 50 cents and that there were 100 armed men standing by to back up the price.

Cotton counted only about 50 to 75 men. In any case, there were enough to show that Rufus meant business. The drivers paid up.

The second and final battle of the Telos War began when the Coe-Pingree operators petitioned the state legislature to build and operate a canal connecting the Allagash River system with the Penobscot. Because there was only one place where such a canal could be built, they were, in effect, asking the state to take the canal and dam away from Dwinel and give it to them.

As Roberts and the Stricklands had not bothered to obtain a charter for their bit of backwoods engineering and Dwinel had seen no need to bother busy lawmakers with such matter, the Telos dam and canal did not legally exist.

Rufus knew he was on shaky ground, but he maintained that the law of eminent domain did not give the state authority to seize his personal property and give it to another to operate for his private gain.

The Maine State Legislature listened to both sides of the issue for a number of days and in due course passed down its judgment. It passed two acts. The first permitted Dwinel to incorporate under a state charter and set the toll at 20 cents per 1000 board feet. A deadline was set for acceptance. If Rufus refused to agree by the deadline, the second act would take effect. This act would empower Coe-Pingree to organize and incorporate the Lake Telos & Webster Pond Sluicing Company, opening the canal to free access.

Old Rufus was caught between a rock and a hard place, and he knew it. He capitulated. The Telos War ended without bloodshed.

Logging of the Allagash headwaters began in earnest. The enormity of the operation boggles the mind when one considers that, except for the power of flowing water to move the logs, everything operated by muscle power, either of man or beast.

Maine State Library collection.

An early 1900s view of Chamberlain Farm. The steam towship Dugan is pulled up on shore.

In 1846 Eben Coe started to clear what was to become Chamberlain Farm. The farm would eventually become an enormous agricultural undertaking of more than 600 acres of cleared land, serving as supply base of lumbering operations for more than 80 years.

The problems involved in toting supplies more than 150 miles into the wilderness were staggering. Anything that could be produced on the farm was money in the pockets of the owners and the operators. Chamberlain Farm raised large numbers of cattle and sheep, as well as potatoes, grain, and vegetables. A sizeable orchard was established, and in time fresh apples were available for woods cooks, who baked hundreds of apple pies to feed hungry lumberjacks.

In one year when the farm was in full operation, records show that it employed seven men and produced 700 bushels of oats and 75 tons of hay.

The buildings at the farm were enormous. They were constructed of eight-inch-squared timbers laid up log-cabin style. No doubt the number and size of these buildings increased through the years. Henry David Thoreau described the farm during his 1857 visit as "two or three log buildings close together." The buildings were later shingled and from the outside looked much like the farm buildings 150 miles to the south.

All of the woods work—the cutting, yarding, and landing—was done in winter. This meant that in the spring, when the drive went down the river the draft animals, mostly oxen at first, but later horses, were left without anything to do until the following cutting season. These wilderness farms, like Chamberlain, provided the necessary work to keep them in shape, and the pasture to keep them fed. In turn, they provided the farm with the power it needed for production.

Chamberlain Farm was the supply base for the whole Allagash headwaters operations, serving not only the Coe-Pingree operators, but

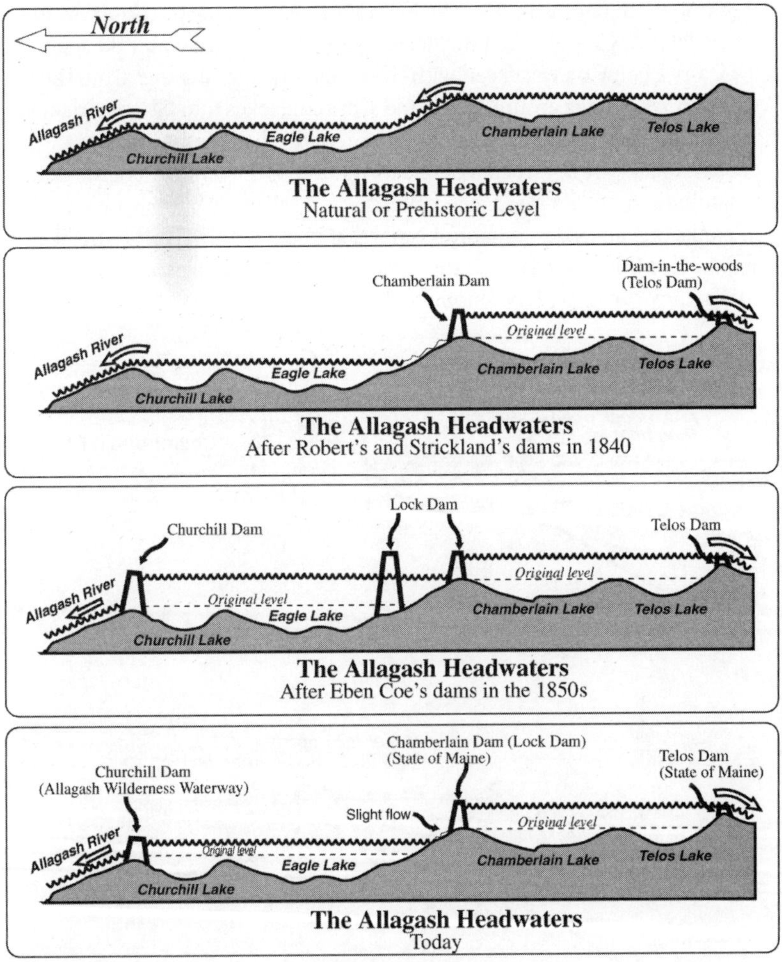

Changes wrought by the Allagash Dams over the years. Most of Chamberlain's water has flowed south since about 1840.

other smaller ones as well. Without this nearby depot of food and supplies, the large outfits could not have operated at a profit, and the smaller ones could not have functioned at all.

Before hearings of a legislative committee, Shepard Boody testified that logs could be moved from the Eagle Lake level to Chamberlain level by hauling them overland, by pulling them up a slip, or by means of a lock and canal system. Eventually all of Boody's suggestions would be used, but he would have been amazed at some of the ingenius methods that were devised later.

Chamberlain dam was originally without gates. Its only purpose was to raise the level of Chamberlain the necessary 11 feet to turn water through Telos Dam. Eben Coe was faced with the problem of getting logs from the Coe-Pingree holdings around Eagle and Churchill lakes into Chamberlain so they could be floated through Telos Cut and on to the Penobscot Boom.

His practical eye saw the favorable formation of the land where the outlet of Chamberlain flowed into Eagle Lake. It was a relatively narrow ravine with high banks on each side—a natural canal that needed only water to make it function. Coe built two dams at the outlet of Chamberlain, each with locks like those any canal used for shipping.

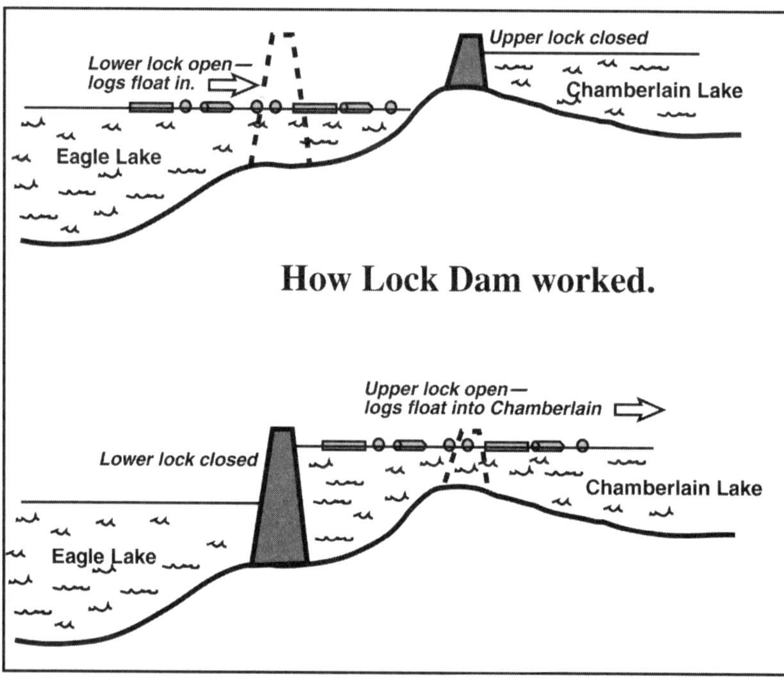

How Lock Dam worked.

While these dams were being constructed, a third dam was built at the outlet of Churchill Lake, the beginning of the Allagash River. As Churchill and Eagle Lakes were at one level, connected by a thoroughfare much the same as Chamberlain and Telos, this dam would raise the level of both lakes.

Like most everything else that Ebenezer Coe undertook, the lock dam concept was successful. The logs from the Eagle-Churchill area were floated into the canal flooded by Churchill Dam. When the waterway was full, the lock of the lower dam was closed. Next, the dam on the Chamberlain end of the canal was opened, allowing the water level in the canal to rise to that of

Maine State Library Collection.

An early photo of Lock Dam.

Chamberlain Lake. From here on the logs followed the route that millions of board feet of Chamberlain logs had taken to the Penobscot Boom.

For the most part, peace reigned throughout the Allagash headwaters during the years following the Telos War. But there were exceptions, especially downriver.

New Brunswick lumbermen deeply resented the diversion of Allagash water south into the Penobscot. The water from the huge Chamberlain Lake was important for driving their logs as well as for producing power farther down the St. John River.

When Aroostook was settled, Maine lumbermen drove logs down the Allagash to mills on the Maine side of the St. John. The resentment of the New Brunswickers over the loss of water was contagious. The whole thing was bound to erupt in violence sooner or later. On one occasion when the Allagash drive was having hard going for lack of water, a party of armed men appeared at the Chamberlain dam with a bag of dynamite. They drove the dam tender away and proceeded to blast a hole in the gates, allowing Chamberlain water to flow toward their stranded logs.

The resourceful dam tender was not one to stand by and watch the water he had carefully and diligently saved for his employers be stolen by a bunch of upstarts from the north. He immediately made his way to the head of the lake, where a boom of logs was being prepared for towing to the cut. The boom was quickly towed to Lock Dam, and the logs used effectively to close the hole. No one knows whether the Allagash men ever received enough water to refloat their logs.

Stealing water—or recovering one's own water, depending on the point of view—was done openly. No action was ever taken against the perpetrators. These lawless acts were considered part of the hazards of lumbering in this wilderness called Allagash.

Lock Dam operated successfully until the dam at Churchill went out. New Brunswick lumbermen were accused of blasting the dam, but it was never proven. Since Churchill Dam was a continuing problem to maintain, their guilt remains in doubt. At any rate, when Churchill Dam was gone, Lock Dam was inoperative. But, since the lion's share of the timber in the Eagle-Churchill area had already been floated south to Bangor, no attempt was made to restore the system. Fifty years would pass before another of Boody's suggestions would be put to work using a technology that didn't even exist in his time.

[CHAPTER TEN]

MUSHRAT MURPHY

Mushrat Murphy was a master of his craft. He knew it. He was fiercely proud of it. No one who worked with him ever doubted his skill once they saw him in action. No one, that is, except that cussed Chipper Costigan! Chipper was the reason that Mushrat was where he was instead of at some downriver lumber camp where a good Penobscot man had ought to be.

Mushrat was standing in a heavy bateau loaded with supplies. He and five other woods workers were poling the heavy boat up a small stream that Mushrat had been told was called Umbazooksus. They would soon reach a small pond by the same name. The man supplying this information labored at his pole directly across from Mushrat. He went on to say that there was a carry ahead that would take them into a little bog-hole called Mud Pond, and then another carry into the largest of the Allagash lakes, Chamberlain. Sometimes this last carry was made easier when there was enough water to float the bateaux through the tiny stream.

Mushrat was not really interested in this information. He kept thinking that the only reason for his being this far upriver was because of that no-good Chipper Costigan. Costigan had been walking boss on the job where Mushrat worked the previous winter. The two men had never had any love lost between them, but mutual respect for the other's ability had kept them working in a relatively harmonious manner. At least up to a point.

Mushrat was a chopper! He prided himself on being the best damn chopper on the Penobscot, and he was quick to prove it to anyone who doubted it. His skill with an axe was legend, and when he pointed at a spot on the ground and said, "A tree will fall there," you could damn well believe that was where a tree would land.

In a woods crew, the head chopper is second only to the walking boss, and the two have to work closely together. The chopping crew consists of three men: the head chopper, the second chopper, and a sled tender. The head chopper and the walking boss work together to decide what is to be done on a particular strip and how it is to be done.

The trouble between Mushrat and Chipper Costigan started in the winter of 1886. Mushrat was working his strip with his crew. They were laying down more trees than any other crew in the camp, which was nothing unusual. Costigan didn't visit Murphy's strip very often because he knew from past experience the crew would do the job right, and besides, he was not overly fond of the cocky little banty rooster of a head chopper who was known as Mushrat Murphy. The farther he stayed from him, the better.

In February the weather was the coldest the men had seen in a long time. In Mushrat's strip were a number of large spruce trees that would make choice logs. But Mushrat explained to his men that they would not cut those trees now because they were frozen solid and would surely split when felled. The crew continued with their work until Chipper paid them a visit. Then all hell broke loose.

Chipper stormed into the clearing where Mushrat and his crew were working and demanded to know why in damnation those spruce weren't cut and hauled to the yard. All of this was said with a stream of expletives that "wan't fit fer perlite comp'ny."

Mushrat replied, "If'n you had a ounce of brains in that thick skull of your'n you'd know they'd split in this cold. We'll leave 'em till spring."

"We won't leave 'em till spring," bellowed Chipper. "If you and those two birds working with you were worth the salt you sprinkle on your beans you cud git them trees to the ground 'ithout splitten 'em. I want them trees cut, and now!"

On purpose or by accident, Costigan had succeeded in reaching Mushrat's most vulnerable spot—his pride in himself and his crew. But, as Chipper stalked out of the chopping Mushrat yelled after him, "If'n you want split logs, you'll git 'em." Mushrat knew he had been baited into a trap from which there was no escape. He had to cut the logs as Costigan had ordered, but when the logs split, as he was sure they would, it would be no one's fault but the head chopper's. Especially as there were no witnesses to the exchange of words between them except his own crew.

Mushrat took his crew back to the great spruces. Though he would have liked to split the trunks in half and throw them in Costigan's face, he was too much of a proud lumberman to do anything but his best to preserve the giant logs. He and his second chopper laid down a bed of small trees for each spruce to fall upon. Then they cut the trees. One by one the giant frozen trunks fell to the ground, and one by one they split open with a resounding crack just as Mushrat had predicted.

That night when the crew returned to camp they heard Costigan talking about how Mushrat and his crew had ruined eight or 10 thousand feet of choice spruce timber by felling it when it was frozen. The other crews were quick to believe anything derogatory about Mushrat's crew, which consistently cut and yarded more timber than any of theirs.

Mushrat was so mad he couldn't talk. He turned and stormed into the time keeper's office and growled, "I'm maken'er out, gimme my time." The time keeper, only mildly surprised because hot-tempered lumbermen often approached him in this way, quietly figured what Mushrat had coming, deducted what he owed the wangan, and handed him his wages.

This was how Mushrat Murphy came to be poling a bateau up Umbazooksus Stream. "All hell would freeze over before he would go back into the woods anywhere that he had the slightest chance of running into that no-good Chipper Costigan again," he had vowed. He knew that with his skill with an axe he would quickly establish a reputation in this new lumber country, but he seethed inside that he was being forced to do so.

The heavy bateau ground to a halt on the shallow riffles where the water flowed from Umbazooksus Lake. Here Mushrat and the others picked up the usual loads and carried them to the lake shore above. Last to be carried up was the bateau itself, which was reloaded and paddled across the pond to the beginning of the trail called Mud Pond Carry.

Here on the northwest shore of Umbazooksus they were met by a man named Ansel Smith, who operated a carrying service with a horse and wagon. Smith, a farmer down on Chesuncook, lived in a small cabin near the lake and operated this little carrying service as a source of income. The men loaded the bateau and supplies on the wagon and it bumped over the muddy carry trail to Mud Pond, the crew strung out behind it.

At Mud Pond the bateau was reloaded and paddled across to where a small stream flowed westward. Here a small dam had been built to store water to float bateaux down the little canal-like stream. The crew walked alongside the bateau, lifting now and then when it snubbed up on a rock. Before long Mushrat could see the light sky of an open lake ahead, and he knew this must be the lake called Chamberlain. Actually, this was only a small cove of the lake, but the crew soon paddled through the narrow opening, and the huge lake opened up before them.

Directly across the lake from where they emerged from Mud Cove was a farm! A huge farm, with acres and acres of cleared fields and more buildings than Mushrat could count. The man sitting next to him in the bateaux told him it was Chamberlain Farm, their destination. From there the men would be assigned to camps and moved out for their winter's work.

As their bateau approached Chamberlain Farm, Mushrat was amazed at the number of men milling around. He could see tents set up to provide temporary shelter for the men, and farther up the slope in pastures behind the buildings were what must have been hundreds of oxen and horses. Mushrat was still taking in this beehive of activity when the bateau scraped to a stop on the gravel beach.

Pulled up on the shore of the lake near where they landed was a large raft, well known to log drivers everywhere in Maine as the headworks. The men in the bateau got out and stretched their legs and hung around the headworks, waiting to be told what to do next. Mushrat recalled the long hours he had spent working on these cussed things down on Chesuncook.

The headworks was made of logs, two or three layers of them, laid crossways over each other, on which was placed a crude capstan usually made of a stump or butt of a large log drilled in the center of the base to sit over a vertical pin and, around the top, over hand-spikes. The boss usually placed a piece of salt pork on the center pin so things would turn a little more easily.

Mushrat's memories of using this terrible apparatus were not pleasant. To tow a boom of logs the length of Chesuncook took three days. Sometimes they worked nights if an opposing wind blew by day. Mushrat judged by the size of this lake that the time men spent sweating on the headworks here was at least that long.

Maine State Library collection.

The Headworks. Before motorized towboats, this was the way logs were moved from one end of the lake to the other.

A hawser, almost as large as Mushrat's wrist, was wound around the capstan, and to the end of it was attached a 300-pound anchor. The unfortunate crew of the headworks would row the anchor out to the full extent of the rope, drop it, and row back to the raft. There they would wind the large spool by straining against the handspikes as they trudged around and around on the raft, drawing the rope and its attached boom of logs slowly toward the anchor and, eventually, the foot of the lake. When the anchor was under the raft they would draw it up, put in into the bateau, and start over again.

Sometimes, if wind conditions favored them, sails and oars would be added and the headworks crew was treated to a relatively easy trip to the foot of the lake. But, sometimes, in spite of all precautions, the boom would break and the hours spent trying to collect and enclose drifting logs turned a dream trip into a nightmare. Mushrat had heard rumors of steamboats being built to take over this job of boom towing, but he had never seen one.

Mushrat sat silently staring at the splintery path the men's caulked boots had made on this headworks when someone yelled for him and the others to follow to the clerk's office. When they arrived, Mushrat lined up with the others.

When Mushrat's turn came the clerk asked sharply, "Name?"

"Murphy," Mushrat answered.

"First name?" the clerk said, somewhat peeved.

"Mushrat," he answered.

The man behind Mushrat snickered.

"I don't carry anyone on my books with the name of Mushrat, what's your real name?" the clerk asked with exaggerated weariness.

Silence! Mushrat could feel the men behind him leaning forward to hear.

"I need your real name; I can sit here as long as you can stand there," the clerk said.

Mushrat stood silent for a moment and then leaned forward inches from the clerk's face and hissed, "Terrance."

He started to straighten up when the clerk said, "What's that again?"

Mushrat snapped back down and, with his eyes boring into those of the clerk, he repeated, "Terrance, you friggin little weasel, and if you repeat it I'll stuff that precious book of your'n down your scrawny throat."

The clerk swallowed and recorded the information. Without looking up he asked, "Are you a river driver; will you work the drive in the spring? "

"Do you think I got the name o' Mushrat from sittin on my arse behind a desk?" Mushrat replied.

"River driver," wrote the clerk. "You'll be going to Bill Moody's camp. You will find him or his foreman out in the yard somewhere."

Mushrat turned and left the desk. He wasn't worried about the men behind him reading his name on the clerk's roster, because he knew they couldn't read any more than he could. He wandered about the area with his precious axe in one hand and his small carpetbag with the few personal belongings he owned in the other. Here and there he saw a face that was vaguely familiar. Probably someone he had seen in a saloon or brothel in Bangor, but no one he really knew.

Woods workers had a definite social strata, and Mushrat could easily pick out groups of teamsters, swampers, choppers, walking bosses, and a few Johnny-come-latelys who didn't know what they were. He sauntered by one small group paying little attention until he heard a familiar growling voice. He could not believe his ears. He dreaded turning around to look, but he knew there could be no mistake. It was Chipper Costigan! No doubt about it. There talking with four or five other bosses was Costigan. He was more or less side-to and had not seen Mushrat.

The smouldering resentment of the past months welled up in Mushrat. He was overcome with a fury he had never felt before. He'd been madder'n hell in all manner of brawls, but this was different. He dropped his carpet bag and carefully laid his axe on top of it. "Costigan!" he yelled at the top of his voice.

Chipper Costigan turned to face Mushrat. A mean smile of recognition spread across his face. He hunched into a fighting stance as Mushrat hurled himself toward him. He was going to enjoy this, he thought as he watched the 140-pound man rush toward him.

Mushrat continued to run toward Costigan. To the men in the yard who had just turned to see what the noise was all about, he was just a blur. When Mushrat was about three paces from his target he jumped into the air and swung his feet upward toward the 200-pound man. As he left the ground Mushrat doubled up his legs and, at exactly the right instant, drove them outward like a released coil spring. One of his heavy leather spiked boots caught Chipper on the side of the head, the other full in the chest. Costigan flew backward, fell, and only moaned. Mushrat stood over him ready to bite, gouge, spike, kick, and punch until the heavier man beat him senseless, which he had no doubt would happen. He didn't care.

But Costigan had all of the fight knocked out of him. He groaned, rolled over on his stomach, and rubbed the side of his head where red spots marked each puncture of Mushrat's spiked boots. There was another set of holes on his chest hidden by his shirt.

Mushrat felt good. Relieved. He walked back, picked up his things, and waited to be fired. Brawls like this were commonplace behavior on Exchange Street, but not up here in the woods, particularly not with bosses. He watched the crowd gather around Costigan and he knew by their actions that he would be all right. He would survive. Then he saw a tall, slender man approaching with a faint trace of a smile on his face. "Well, here it comes," thought Mushrat, "and it looks like he's goin to enjoy it."

"You must be Murphy," the man said. "Name's Moody. Hear tell yer goin to be part of my crew. If ya c'n handle that axe as good as them corked boots yer wearin, we're goin to git on fine. Man, ya really put tha boots to 'im. He'll have a case of logger's smallpox for a while. Never liked that blow-hard from the time he showed up here. Glad he's goin to be workin' t'other side of the lake."

Moody stuck out a calloused hand, and Mushrat shook it gratefully. "Come on," said Moody, "I'll find ya a place to flop fer tha night. We're movin out in the mornin. Yer the last of my men for now. Rest'll be along after freeze-up when we've got the camp built, roads swamped, and streams cleared. Ever built a camp?"

Mushrat nodded and followed the woods boss. For the first time he started to feel good about being here. It wasn't such bad-looking country at that, he had to admit to himself.

Bill Moody's contract called for him to cut an area in the upper reaches of Ellis Brook. He did not underestimate the amount of work that had to be done before the freeze-up when the cutting would begin. This area had never been cut before. Oh, the big pines near the water ways had been harvested, but a systematic cutting of the giant spruce and what remained of the pine had never been done.

Ellis Brook had not been driven before. This meant the centuries of accumulated driftwood and fallen trees that covered the streambed would have to be removed. Much of the ground around the stream bed would be so swampy that horses could not be used. The men would do the work by hand. Tree by tree, stick by stick, the obstructions would have to be lifted out and put back far enough from the shore so they wouldn't be washed in again by high water. Islands and shoals would have to be dug away. Stumps embedded deep in the mud had to be grubbed out; embankments had to be made at sharp bends. Sunken trees in the bottom of the stream would have to be cut and worked loose in the beds they had lain in for years. Then a chain would be fastened around them and reluctant horses forced into the freezing water to haul them away. The men would be in the cold water until they cleared the channel and built a driving dam to store water for the spring drive.

The buildings that would house the men and animals were the first order of business. The choppers went to work cutting trees for the needed logs. Mushrat quickly showed his skill with his axe as he laid the trees down in perfect order for the teamsters to haul to the campsite. His stumps were as smooth as marble, and the chips around where he had been chopping were nearly as big as a dinner plate.

Moody and his foreman were quick to notice his skill and soon had him notching and fitting logs for the camps. Without any official appointment or mention of it, Mushrat was put in charge of the construction. He insisted on the same precision in building that he exhibited with his axe.

The first step in building a camp was to clear away the flammable duff or forest litter so the floor, at least, would be fireproof. Spruce logs were then notched and laid up for the walls. The camp would be square, about 20 by 20 feet. The logs were laid up and fitted as closely as possible and the cracks between them chinked with moss or peat driven in with a wedge and maul.

The back wall of the camp was about three feet high. The side walls, which supported the horizontal log roof rafters, were cut back so the camp had a shed roof of about eight feet high at the front wall.

A large rectangular opening was left in the roof for a smoke hole. A log chimney supported by roof rafters was built around this hole. The roof was made by laying "splits," or shakes, split from bolts of cedar three or four feet long. The splits were overlapped so they would shed water. They were held in place by long poles and then covered with a thick layer of fir or spruce boughs. The roof would not only shed water, but when the boughs filled with snow in winter it would provide a snug covering as well.

The only opening in the walls was near the front corner, where a door made of splits was hung on wooden hinges and fastened with a wooden latch.

All the men slept in a community bed that occupied the entire low side of the camp. The low wall of the building served as headboard, a small log as footboard. The ground between the footboard and the wall was covered with a thick layer of balsam fir boughs that offered a soft, fragrant, warm mattress. Bedclothes consisted of blankets or quilted spreads sewn together to a length that covered the entire berth. When the camp was filled to capacity the tightly-packed sleepers were the model of drowsy cooperation. When one man decided to roll over or otherwise shift his position, all the others had to rearrange themselves until the entire line was once again comfortable.

Running the entire length of the camp at the foot of the bunk was a half log, hewn smooth on the top. It was known as the "deacon's seat," though no one seems to know why today.

When winter snow piled up around the camp, a fire built with four-foot logs kept the men warm and dry. At times the wind blew the smoke back down the smoke hole, sending everyone outside coughing and hacking. Sometimes the roof caught fire when the fire flared up a little too high and hot cinders ignited the dry cedar splits. However, there was always plenty of snow and water on hand, and plenty of people to help put out the fire.

The shelters for the animals were much the same as the men's camps, but a little more care was taken to ensure the comfort of these important beasts. After all, if a man got cold or wet he could move to the fire. The oxen and horses could not. Dirt floors were considered good enough for the men's camp. For the animals they built a floor of small logs hewn on one side to

Walker Transportation Museum collection.

In the early days of Maine, lumbering oxen were the source of power that moved the logs from the woods to the water's edge.

make a flat surface. Teamsters were diligent in caring for their animals, always putting the comfort of the ox or horse before their own.

Mushrat had never been happier than he was that winter in the woods of the Allagash. He was made a head chopper, and he couldn't have picked a better crew to work. His second chopper was a young man named Danny Arnold. Arnold made no secret of his admiration for the head chopper. He hung on Mushrat's every word. The attentiveness and esteem Arnold displayed prompted Mushrat to take extra pains to tutor his assistant.

Working with them was Manley Doan, who was as dedicated to his team of oxen as Arnold was to Mushrat. Mushrat had often seen him kneeling beside the animals in sub-zero weather, melting the ice from their legs with his bare hands.

The walking boss Mushrat worked for was fair. As long as a man did his job, Henry "Hardstick" Blodgett gave him no trouble. Blodgett and Mushrat were usually of like mind about how a particular strip should be cut. When they did disagree, Mushrat didn't mind deferring to him, because the man knew his job. By the same token, Hardstick listened to his subordinates. If they made sense, he'd do it their way.

But it was Bill Moody whom Mushrat admired most. The contractor visited the camp often, and the two men got to know each other well. Moody had started out in the woods as a teenager, working as a cookee, a cook's helper. Then he worked as swamper and continued up the strata of woods workers until he had charge of several camps, all known as Moody's camps.

Two choppers at work on a white pine. The axes are seen as a blur, but the notch of the near chopper is clearly seen.

What impressed Mushrat about the man was how well he fit in and got along with the other men, and how he didn't mind helping with any job that needed doing. Like the day he showed up at dinnertime and the cookee was sick. Moody picked up the food and headed out to deliver it to the crews. He got to Mushrat's strip last and sat down with the men while they ate. He told a story that day that Mushrat thought was the funniest story he'd ever heard.

It seems that when Moody was a young man he had become a pretty good second chopper. So he went to a camp looking for a job where he had heard one was available. The walking boss was a big blustery fellow who had been a first-rate chopper. He took great pride in that fact and even greater pride in his axe, which he always had with him, its edge always keenly honed.

"So you think you're a chopper eh?" the boss asked Moody. "Before I hire any choppers, I want to see how they handle an axe. There's a pile of wood outside the door. Go split it, and we'll see how good you are."

"Ain't got my axe with me, " replied the young Moody. He had left his axe and bag on the supply wagon he had ridden into camp.

"Ain't got your axe with you," roared the boss, "what kind of chopper do you think you are, trying to get a job without an axe? Here, use mine."

"T'won't be hung right for me," replied Moody. Every axeman attached the handle to the axe head to suit himself. This was called "hanging the axe."

The boss laughed at him and said, "If you can't split a little wood with someone else's axe, I doubt if you can chop down a tree with your own. "

By now, Moody was so mad he'd made up his mind he didn't want to work here anyway. But he grabbed the boss's axe and went out to the woodpile. The boss followed him and pointed to the small pile of wood and then to a large rock. "There's the wood and there's the chopping block," he said, pointing to the piece of granite.

Moody set each piece of wood on the rock and deftly split them one by one, giving the axe a twist as it struck the wood. The pieces flew apart. The axe head never came close to the rock. Moody had the pile split in five minutes. Then he reared back and drove the axe onto the granite rock with one forceful swoop, causing a horrified reaction from the boss.

"Why in the hell did you do that?" the boss screamed.

"I always leave my axe in the choppin block when I'm done splittin," Moody said quietly over his shoulder as he walked away to collect his belongings.

This was the kind of story that could really tickle a chopper's funny bone. Mushrat chuckled to himself every time he thought about it that winter. Mushrat thought about other things that winter too. Mostly about himself and what was to become of him.

Before coming to work for Bill Moody, he'd never thought of being anything but a chopper. But now, realizing that this man he admired had worked his way up from the bottom, he asked himself, "Why not me? " If he continued to work as a chopper, he'd only go back down the ladder when he became too old to swing an axe. Many a lumber camp had an old cookee who had once taken great pride in his ability to swing his gleaming axe.

Mushrat already knew as much as most walking bosses, and a lot more than some he'd known. He always handled his crew of men well. How different could it be to handle two or three crews rather than two or three men? He knew he had things to learn, but for the first time in his life he felt that he had a purpose other than just getting through the winter so he could get to Bangor and spend all his money on women and whiskey.

Bill Moody was the kind of man he could talk to about such things. He'd do it, too. He'd ask Bill if there was something for him this summer so's he could learn. And, if there wasn't a boss's job for him next year, then maybe the next. But he'd do it.

"Someday," he thought, "maybe there'll be Murphy's Camps. Wouldn't that be som'thin."

[CHAPTER ELEVEN]

THE TRAMWAY

The turn of the century found the lumbering scene in the Maine woods much changed from the previous 50 years. Not only were the cutting methods changing, but also the tools and living conditions of the men.

Crosscut saws replaced the ax as the primary means of felling trees. This was made possible by the invention of the raker tooth which prevented the saw from binding itself in its own sawdust. The cutting crew consisted of the head chopper, two sawyers (who were still called choppers), a sled tender, and one or two swampers. Each had their duties, but all pitched in and helped one another when their own chores were done.

The head chopper would face notch the tree with his axe. The sawyers would saw the back of the tree while the chopper made sure they sawed evenly across, parallel to his cut. If necessary, he would have them hold back on one side to leave a "bone" that would serve as a pivot to throw a leaning tree one way or another.

The lowly swamper—the greenhorn of the crew, or an old man who could do nothing else—swamped a road so the teamster could get to the log with his horses. When the tree was down he worked at limbing. Some say the daily cut of logs was doubled when woods crews started using the crosscut saw.

The lumber camps, while not luxurious by any standards, were much improved. The shed-roof style buildings gave way to the peaked roof type, and many had one or two glass windows that let in a little daylight. These windows were a luxury, because few men saw the interior of the camp during the day except on Sundays.

The camps, still built of logs, consisted of two buildings end to end separated by 12 to 15 feet. The space between the buildings was called the "dingle." It was covered by a roof and used to store food and supplies. The building on the left, where the men lived and slept, was called the barroom. The building on the right served as cookroom and dining hall.

The barroom usually had a wood floor. It was heated by a large monster of a stove, which was usually manufactured by Wood-Bishop Company of

Bangor. The men still slept in a common bed, but now it was almost always a raised platform. The double bunks didn't come along for another couple of decades.

In earlier years the men stood and ate out of a common pot. Now they sat at a hewn-log table and ate their meals from tin plates. The cook ruled the camp. No one spoke during the meal, an unwritten rule that was passed on through the years. The cook had to get through the meal, clean up, and immediately start preparing another mountain of food for the next meal.

Occasionally some renegade would violate the silence. But he was quickly squelched, unless the cook had failed to please the crew and pressure was being exerted to have him fired. If the men pointedly ignored the no-talking rule, the cook would get the message and either quit or face the prospect of being fired.

The Allagash lumbermen at the beginning of the twentieth century were feeling pressure to speed things up in order to hurry raw materials to the rapidly industrializing country. The lock dam concept of raising logs from Eagle Lake to Chamberlain had worked admirably, but it was slow. The job

Walker Transportation Museum collection.

The upper portion of the tramway. The lower (return) tracks were underneath.

Tim & Sue Caverly collection.

Another view of the tramway. Eagle Lake can be seen in the background.

had to be done faster. Shepard Boody had testified, "haul them overland, pull them up a slip, or use a lock and canal system."

Boody, no doubt envisioned men or animals pulling logs up a wetted trough into the lake above when he recommended the use of a slip. But mechanization was coming to the Allagash.

In 1901, Fred T. Dow of Bangor surveyed the 3,000 feet of land that separated the head of Chamberlain Lake from Eagle Lake. In his report to his employers, Marsh and Ayer, he stated the feasibility of a mechanical tramway that could take logs from Eagle Lake and St. John waters across the hill between the two lakes into Penobscot waters. A mechanical slip! Chamberlain, of course, could now be made to flow either north or south by use of dams. But Bangor men were never going to concede that this lake was anything but Penobscot water.

Terry Harper collection.

In the foreground are the steamship boilers, probably the Dugan's. In the background is the building that housed the tramway machinery. The smokestack rises from the two wood-burning boilers.

Mr. Ayer gave his go-ahead for the project. The contract for all steel castings and cable was awarded to the Taylor Iron and Steel Company of New Jersey. The New Jersey firm was chosen because it was the first in the country to manufacture manganese steel castings.

Moving machinery into the woods was a gigantic undertaking in those days. Tote roads were nothing more than cleared paths through the woods over which teamsters maneuvered their loaded wagons, sometimes jolting over the stumps. As the soft forest floor was compacted by each successive passing wagon, the stumps and rocks seemed to grow and become more and more troublesome. Sometimes the wagon drivers had to strap themselves to their seats to keep from being pitched off.

Nearly all of the tramway machinery was taken across Moosehead Lake to Northeast Carry in the fall of 1901. What was left at Greenville when the 1ake froze over in the fall was taken across the lake with teams by H.N. Bartley. Brown and Wiggin of Patten contracted to move the machinery from Northeast Carry to Eagle Lake. These heavy loads of boilers, heavy cast trucks, and cable were extremely formidable for those days, and the contractor had his troubles.

The cable weighed 14 tons. They wanted to take it into the woods in one piece. They rolled it on two drums and mounted it on skids hauled by horses. Below Smith's halfway camp on the West Branch of the Penobscot, progress of the move ground to a halt. Then a man named Phil Sawyer came forward and solved the problem. Sawyer had been a cable splicer by trade. He offered to cut the cable in half and then splice it back together again at its destination. The two drums were separated, and the rest of the move was accomplished without any serious problem.

On March 1, 1902, Dow, with a small crew, started on the foundations for the boilers and the heavy machinery. They had a small rotary saw at the tram site that was used to saw timber. It was to have been powered by the large tram engine brought in with the machinery, but its crankshaft was broken on the way in. The saw had to be run with a donkey engine until the fall, when a new crankshaft arrived for the main engine. Before the main engine was operable, Mr. Dow had turned the conveyor over by power from the donkey engine. It was late in the fall before the main engine could be used.

The broken main engine was not the only difficulty that had to be overcome before the tramway could be used. Each of the 600 trucks and 600 clamps had to be fastened to the seemingly endless cable by four seven- or eight-inch bolts. When the tramway was loaded with logs the trucks and clamps started to slip. The threads on the bolts had not been cut down far enough to allow the nuts to be tightened sufficiently around the cable. Each of the forty-eight hundred bolts had to be removed, threads cut down with a hand die, then replaced.

A man named Ralph I. Miles worked at the Tramway when it started in 1903. Here is a letter written by him to Mr. David Withee about those times. It gives a vivid picture of what life was like in Maine's north woods very early in the 20th century.*

Dear Mr. Withee:

I have read with much interest the article about the Tramway and the letter you wrote to Mr. Bragg.

The reason I was interested is that on July 1, 1901 as a boy or young man of 20, I left Patten for Chamberlain Farm. We went to Sebois Farm by buckboard and from there with a jumper on what was after [known as] the Eagle Lake Road. The swing camp was not there then. But, 22 miles from Sebois Farm we swung west, crossed the little East Branch at the foot of Red Falls, from there to Sly Brook and so on to the Arm of Chamberlain. We were three days from Sebois Farm reaching Chamberlain Farm as we had to shoe the jumper every day with wooden shoes. I stayed at the Farm until the next spring so you say [see] it was quite a stretch for a young fellow. I, at that time worked for Wiggin and Brown who hauled all the material for the Tramway from Northeast Carry on Moosehead Lake, except [for] the boilers and engine and cable. The Tramway was built in [the] summer of 1902 and started the spring of 1903. I was at the Farm four summers and two winters. Worked for Wiggin and Brown four years. Two winters at the swing camp. I was at the Tramway the day it

* From the Allagash Alliance Group collection

Terry Harper collection.

Building of the steamships Dugan (above, Chamberlain end) and the Marsh (below, Eagle Lake end). Both were built in 1903 by O.A. Harkness, standing at the far right of the lower photo.

Terry Harper collection.

Towships Dugan and Marsh. The Dugan (above) is shown pulled up on shore, probably at Chamberlain Farm. The Marsh looks to be frozen in for the winter.

started. At first it bothered as it was loaded too heavy, but they would put on a few logs [and] as they tipped the height of land those logs helped to pull the others so [in] only a short time it could be loaded full..........

Harkness built two boats for Marsh. One on Eagle Lake and one on Chamberlain. I have been tole [told] that one laid on the shore of Chamberlain at the Farm and rotted there. I remember Harkness and all the old timers very well.

The winter of 04 and 05 I clerked for Ira Carpenter at the Arm of Chamberlain and the wages then were $26.00 a month for the best men such as choppers and sled tenders. Teamsters got $1.00 a day or $30.00 a month. Swampers: $20.00.

At that time Lawler Bros. Were logging at Allagash Lake and Murphy on Telos Lake. I called on Billie Murphy in Old Town about a year before he died. John Kelly who has been dead for a number of years took over after Marsh and used the Log Haulers. Steam.

I have made two copies of the Tramway articles. One for Mr. Bragg and one for myself.

I would like to hear from you.

Yours truly,
Ralph I. Miles

Once the tramway was in operation, it came under the supervision of O.A. Harkness, a mechanical genius. Later Harkness developed an inland fleet of lumbering boats for Great Northern Paper Co. and became known as "The Admiral" at Great Northern. In the November 1927 issue of *The Northern,* Harkness described the tramway this way:

"The steel cable, 1-1/2" in diameter, was 6,000 ft. long, and fastened together so that it was endless and reached from Eagle Lake to Chamberlain Lake. At intervals of 10 ft. the trucks were clamped on. These trucks consisted of a steel saddle on which the log rested, and two 11" wheels which ran on steel rails 22" apart. There were two tracks, one above the other. The loaded

Terry Harper collection.

One of the early steam-powered Lombard Log Haulers. The engineer rode in the back cabin. The steersman rode in front. The roof over the steersman is not to protect him from rain, but from the sparks from the wood-burning boiler.

Walker Transportation Museum.

A later model Lombard, this one powered by gasoline. Steering was done from the cab in the rear.

one went on the top track and the empty one returned on the lower track. Halfway between the trucks there was a steel clamp. Both the clamp and the truck fitted into the sprocket wheel, which was 9 feet in diameter, situated at the Chamberlain end of the tramway. This sprocket wheel made nine revolutions per minute which made the log travel at the rate of 250 feet per minute. The sprocket wheel geared to a Westinghouse Compound Engine, designed especially for electric light plants. The cylinders were 12" and 24" with a 14" stroke. The engine made 255 revolutions per minute with 100 pounds of steam. Wood was used for fuel for the two boilers furnishing steam for the engine. It took a lot of power to start the machinery moving, but it rolled easily once it was in motion.

"The tramway was used for six seasons. It was under my care for the entire time. We averaged about 500,000 feet, board measure, for each day operating, the day beginning at 4 A.M. and ending about 8 P.M. This tramway was always an excellent piece of machinery and very efficient in doing the work for which it was designed. The total amount of lumber taken over the tramway was about one hundred million feet."

The tramway ran right into the water of Eagle Lake and turned a submerged wheel. Here men with caulked boots and pick poles pushed and shoved the logs into place so the trucks could pick up the logs as they came out of the water. On the Chamberlain end a spur wheel picked up the logs and pushed them down a slip into the lake.

Practical working men are forever plagued with the theorizers from the "front office." Herbert Marsh, supervising the loading operations, insisted that the men put one log on each truck. The weight was too much for the cable, and it kept stopping. Harkness, the ostensible straw boss, came from the Chamberlain Lake side to see what was causing the delay. Marsh asked him if there was anything he could do. Harkness answered, "Yes, go back to Bangor and let us handle the damn logs the way they ought to be handled."

O. A. Harkness had not come to Allagash country intending to run the tramway. He left Bangor on January 1, 1903, and headed for the tram site to build a tow boat to tow the logs across Chamberlain, down through Telos into East Branch waters. The boat, named the George A. Dugan, was 71 feet long with a 20-foot beam. It had two boilers and two engines. It was built at the Chamberlain end of the tramway and completed in May 1903. In April of that year F. W. Ayer had written to Harkness to ask him to take overall charge of the tramway. He agreed.

The Dugan was not the only steamer built then. The H. W. Marsh was built on Eagle Lake in 1903 and towed booms to the tramway for five summers. When her work was complete on Eagle Lake, the Marsh was taken to Chamberlain Lake to help the Dugan tow booms to Telos. The Marsh was 91 feet long, had a 25-foot beam, and drew four feet of water. She was equipped with a 150-horsepower cross compound engine with two vertical wood-burning boilers.

Both the Dugan and the Marsh were used for about 10 years. The useful life of the H. W. Marsh, a side-wheeler, ended with her broken in two on the shore of the lake at Chamberlain Farm.

One fall the Marsh was partly frozen into the lake. To protect the engine, her stern was cut off just behind the paddle box and her bow pulled up on shore. When the ice went out in the spring, the stern floated away. For years pieces of the stern washed up on the gravel beaches and were bleached by the sun along the miles and miles of Chamberlain shoreline.

The 100 million feet of timber that traveled over the tramway during its six seasons of operation was not, by a long shot, all of the Allagash timber to move across the height of land into Chamberlain. Boody's prophecies became reality when a man from Waterville, Maine, invented what became known as the Lombard Log Hauler. This mechanical monster quickly proved more efficient than the tramway and was directly responsible for the tramway's demise in 1909.

The Lombards were the first tracked vehicles—forerunners of the bulldozer and the tank. They were a winter vehicle used on iced roads. Huge sprinkler sleds, each holding 13 thousand gallons of water, prowled these woods roads through the night, preparing a glassy surface for the next day's hauling.

The crew of the log hauler consisted of an engineer, a fireman, and a driver or steersman. A conductor had overall responsibility for the train of seven to ten sleds. The conductor rode in relative safety in the rear of the sled and signaled the engineer with a bell, which was rung by pulling a rope strung between the log hauler and the conductor's position at the rear. Pity the poor driver! The earliest log haulers used a pair of horses for steering. But the horses were quickly replaced by a steering wheel. The driver sat at the front of this nearly 150 tons of wood and machine and daily risked life and limb as the heavy machine careened down ice inclines, often out of control. All he could do was keep the runners pointed in the direction of the road and hope for the best. Then the wind would sometimes blow glowing cinders from the huge smoke stack onto his clothing, and he might find himself wearing a burning hat or coat. There he sat with a death grip on the huge iron steering wheel, and as he approached each downgrade he hoped he would survive it so he could search for a new line of work. No use jumping. If he were to land on the icy road with ten heavily-loaded sleds slewing and sliding around, he wouldn't have a chance. His exploits were the topic of many a conversation along the deacon's seat at the lumber camps. Later, when gasoline Lombards came along, the driver became driver-engineer (or vice-versa) and sat in a cab behind the engine.

The machines were very expensive for the time, about five thousand dollars FOB. Waterville, Maine, and the men in the woods who kept them running must have been marvels of mechanical make-do-itiveness. Many parts were made of cast iron, which was state-of-the-art for the time but of low quality by today's standards. In the extremely cold Maine winter these cast iron parts would break. They were a constant source of aggravation to the backwoods mechanics. This problem became so acute with some of the track parts that Lombard resorted to making them from more expensive machined steel. Getting replacement parts often took weeks. Anything that could be done to keep things going was tried. Many an old Lombard today wears the evidence of the inventiveness of some woods mechanic—piece of wire, wood, leather, or scrap metal. These may have been a temporary fix. But some temporary parts worked as well as the parts they replaced, so they became permanent.

Not only did the job titles of the crews of the Lombard "trains" have a ring of the railroad to them, but their scheduling did as well. A dispatcher controlled traffic on the icy roads. The length of a road was referred to as a two turn-out road or a four turn-out road, depending on the number of places provided for one train to "turn out" and let another pass. A telephone line was strung along the route with as many as ten telephones on it. As one of the big machines passed, word was sent to the main dispatcher.

The Lombard greatly increased the logistics problems of the Allagash lumberers. Not only did the giant machines require fuel, but roads had to be improved for them to haul over, bridges and trestles had to be stronger to bear the increased weight, and the sleds used to haul the logs had to be larger and stronger. The expensive machine best served its owners when it was kept constantly on the move. So two or three sets of sleds were used. One set was loaded while the log hauler made a trip with those already loaded. Some larger operations even had go-back roads so the returning machines would not have to turn out and wait while another passed.

Alvin Lombard ceased to manufacture the steam log haulers in 1917, but their somewhat brief tenure in the Allagash marked the beginning of a new era in the woods. Like much of the other machinery used in the logging of this wild country, the Lombard Steam Log Haulers were abandoned where they were last used. Many were left at the woods depot where they sat waiting for parts, and many others in some unlikely out-of-the-way spot where they broke down for the last time.

[CHAPTER TWELVE]

JILL-POKES IN THE WILLYWAGS

James McKinnon was already awake when he heard the cookee clomp across the dingle and into the bar-room to call the men to breakfast. He had been lying there awake listening to the changes that the warm spring weather was bringing to the woods around the camp where he had lived and worked for the past winter. Each of the bunks in the barroom were double bunks. Sharing his was his friend, Brian McGargle, or Mac as everyone called him. Mac wouldn't be awake, even after the cookee gave his rasping call. Jim would literally have to pull him to his feet before he would get up and get going.

Once awake, Mac was a dynamo of happy-go-lucky, carefree energy, but he liked his sleep in the morning. This habit had cost him several jobs over the years as various walking bosses became fed up with him strolling out for work long after the sun was up. Jim didn't mind the daily struggle, even though Mac always grumbled and had some choice words until he was fully awake.

The camp was at the mouth of North Twin Brook on Spider Lake. This morning the water running over the dam and in the stream sounded louder and clearer than usual. *Maybe today we'll break in the landing and get on our way*, Jim thought.

Jim and Mac had spent the winter cutting from this camp, and now they would become drivers as the logs were driven down the brook to Churchill Lake. Most of the driving crew in the camp were, like them, men who had spent the winter here. Some had elected to leave by spring thaw when the cutting was brought to a halt. They were replaced by a few men who arrived with a "two bushel trunk"—a grain bag, in which they carried no more than a handful of extra clothes and a few other belongings.

Jim was more serious about life than his devil-may-care friend Mac. He had worked 12 of his 26 years in the woods, starting out as a cookee when he was just a boy of 14. He liked the work and couldn't imagine earning a living any other way. Unlike many of the men who worked with him, he could read and write; he had gone to school through the eighth grade down in Fort Kent.

There were probably plenty of jobs he could get down home, but he didn't care. He liked the woods and was going to live his life here.

The camp they were in was one of Murphy's camps. There were several others on the east side of Churchill and Eagle Lakes; all of them were owned by M. T. Murphy, who was a jobber for King Lacroix. McKinnon greatly admired the little white-haired Murphy who, like him, had started out in the woods as a cookee.

M. T., as everyone called him, was a man who tended business. He was frequently seen at all his camps, axe in hand, and he made a point of knowing and talking to all his men. He put up with no nonsense from his walking bosses or his men, but he would back them all the way when the situation demanded it. His loyalty was reciprocated, and M. T. was one of the most successful jobbers on the Allagash.

For a man so well liked, a lot of mystery surrounded him. How old was he? Guesses ranged from 50 to 80. What was his name or what did M. T. stand for? One old-timer Jim had known said he used to be called Mushrat, but Jim couldn't imagine anyone calling M. T. Murphy Mushrat! But there was another story about this wiry little man that Jim did believe. The story had it that back in the 1880s or 1890s he had a terrible fight with a man at least twice his size. There were plenty of men in the woods who were twice Murphy's size. The fight was said to have lasted for two hours. Both men were staggering when Murphy delivered the final blow with his caulked boots, knocking his opponent unconscious and then falling across him. Jim figured that with the energy M. T. had today, the tough little man must have been a fireball when he was younger. So a story about his licking a 250-pounder wasn't hard to swallow.

As he had done each morning for the past several days, Jim stopped in the dingle and looked out over the lake. It was still dark out, but he was sure the lake was free of ice for the first time since last November. It never ceased to amaze him how one day the ice was there and the next it was gone. Some said it sank, but he knew better. He'd lived around water all his life and he knew the ice just rotted, honeycombed, and then dissolved, almost all at once it seemed.

The walking boss, Brian Carey, joined him and said, "If the brook's clear we'll break 'er in today and get this show on the road."

This was good news to Jim. After a few days of more or less inactive life around camp, he was ready to go to work again. He eagerly anticipated the drive. Doing something different after a long winter of cutting appealed to him.

The shores of Spider Lake and North Twin Brook were lined with piles of pulp wood and long logs. Both pulp and logs would be driven down the brook to Churchill Lake, but there they would be separated. The pulp would be boomed and hauled down to Eagle Lake, where King Lacroix was just finishing up a railroad that would haul the pulp over land to Umbazooksus Lake. From there it would be driven down the West Branch of the Penobscot to the Great Northern Paper Company mill in Millinocket.

The long logs would be hauled to Churchill Depot, where they would be sluiced through the dam and on down the Allagash and St. John rivers to Canadian mills. This was the drive Jim and Mac would be working. Eventually they would end up in Fort Kent, where Jim would visit with his father and mother for a while before returning to the woods.

By the time the men finished breakfast Carey was back and gave the order to "Break 'er in"—roll the log piles into the water and start the drive. The men picked up their peavies and trooped off to the huge piles of logs and pulp wood. They would spend several more nights at this camp, until the rear of the drive cleared, and then they would move down to Churchill Depot. When they left the depot they would live out of the wangan and sleep in tents.

Drivers were assigned to various jobs according to their skill and experience. The newest men would break in the piles around the lake and then sluice the logs through the gates of the dam, then bring up the rear of the drive. Jim, Mac, and the other river-wise drivers would be stationed at potential trouble spots along the brook to prevent a jam. The men on station were always within sight or hearing of each other so when there was trouble they could quickly call for help, and if a jam was serious, word could be passed upstream to stop the sluicing.

By mid-morning North Twin Brook was flowing full and fast with stored water from the dammed-up Spider Lake, and long and short logs were riding the flood toward Churchill Lake. Jim's station was at a sharp bend in the brook where there was a ledge that projected above the surface, even in this high water. His experienced eye quickly saw the potential trouble from the ledge. He pulled in the first of the long logs that came by and worked them into the eddy above the ledge. Then he arranged these logs so they acted like a wing, diverting the other logs clear of the ledge and into the deep, smooth-flowing water below. Then he had nothing to do except keep his wing in position and relay signals up or down the brook.

By the time the cookee showed up with their second lunch at about 2 p.m. Jim was bored, but not so bored that he hoped for action. Action meant a jam. Jim knew the frantic work that would be needed to release a jam on the flooded brook. There was always an element of danger that he accepted as part of the job, but by the same token he was not eager to be ground under a pile of logs as they rushed downstream. His friend Mac, on the other hand, seemed to thrive on this element of danger and would remain at a jam until the last minute before jumping free.

The first day of driving went without serious incident. No major jams. A few small ones had developed, but the experienced drivers from nearby

This memorial reminds us of just how dangerous the life of a driver was. It reads: *Many friends regret the death of Joe Howell killed here by an explosion of dynamite April 2, 1912.*
There were many graves along Maine's rivers. The usual memorial was to hang the deceased's corked (caulked) boots on a stake or nearby tree.

stations had released them and the sluicing went on without interruption. At dark the gates of the sluicing dam were closed, and when the men on station saw the water dropping they shouldered their peavies and trudged back to camp, where they would have their fourth meal of the day.

The second day on North Twin Brook went well enough until about noon, when Jim heard a call from below. When he looked he saw a huge log slowly tip upright and stand there pointed skyward, entangled in the growing jam of long and short logs. Just below Jim's situation there was a large log "whirling in a roozling hole"—caught in an eddy. He ran down to the shore and leaped onto the log, driving his peavy into the solid wood. His momentum drove the big spruce log into the main current and sent man and log speeding toward the jam below. Just above the pile-up of logs Jim started the log rolling, and in this manner propelled it and himself sideways close enough to shore so he could jump off. Then he pulled the log in so it couldn't join the jam before he went below to help.

The trouble with driving long logs and pulp together is that in a jam, the little four-foot pulp logs wedge themselves in the long logs and make the mass tighter. One look at the jam and Jim knew it was a bad one. He sent one of the younger men back up to the rear to tell them to stop sluicing. Soon there were a dozen or more men at the jam and the work began in earnest. Once in a great while you were lucky enough to find a key log which, when removed, would release the whole mess. Most of the time, though, picking a jam consisted of pulling, lifting, tugging, and rolling one log at a time until enough of them were removed so the rest of the pile would yield to the pressure of the water and the "jam would haul."

When prying and twisting with peavies failed to break a jam, experienced men with axes cut the logs loose, trusting their nimble feet to get them out of the way if the jam hauled. If things really got bad, dynamite was used to blast free a single log. A hole was bored into the log and a half stick of dynamite, with detonator and fuse, was packed into the hole and set off.

Jim and the others worked for an hour or more moving logs. Suddenly, as they rolled one whose end was buried in the pile, the logs beneath them started to move. The men leaped across the moving logs onto the shore, and turned to watch the jam haul. All the men, that is, except Mac McGargle.

This was the moment Mac had been waiting for. With a grin on his face and both hands gripping his peavy, he nimbly jumped from log to log as the jam churned and gained momentum as it moved downstream. Each time he jumped, it seemed to the rest of the crew that there would be nothing for him to land on. But then a log would appear, and his caulked boot would touch it momentarily before he hopped to another log. McGargle kept up this dance until he found a log that had stabilized and was floating more or less level. He landed on this haven in the midst of the churning, splashing, grinding mass of logs, drove in his peavy, and rode the log downstream to his station. The rest of the logs followed McGargle's log like children following the Pied Piper. Jim breathed a sigh of relief, muttered "Damn fool," and turned to go to his station where he would send word to resume sluicing.

Jim knew Mac would never stop his antics. Mac relished the thrill of these rides even more than the notoriety they brought him. Some of the men might disapprove of Mac's penchant to show off, but none could deny his skill on a log. The ability to ride a moving log through fast water, or even propel one in quiet water, was taken for granted by experienced river drivers, but few honed the skill to the degree Mac had.

Sometimes when logs were being driven from small ponds, they would take up all the water, and the gates of the driving dam would have to be closed until the pond filled up again. Spider Lake, however, had plenty of water to see the drive through and, by the end of the seventh day, the sluicing was completed and the rear of the drive started down.

Progress was gauged by the rear of the drive. A drive was not considered complete until all the stray logs were carried, rolled, or dragged back to the water and sent downstream. This "rearing," as it was called, was the least desirable of the river driving jobs, and because it required no special skill, it was assigned to inexperienced men.

When the rear of the Spider Lake drive reached Churchill Lake, the men made their way back to camp for the last time. In the morning they would hand their meager belongings to the cookee so they could be carried with the wangan. At Churchill Lake the logs already had been boomed and were ready for towing. The pulp would go south, the long logs north.

The crew would travel by boat to Churchill Depot, where the men would become part of a larger drive that would end a hundred or more miles down river on the St. John.

Churchill Depot was a hub of activity. Standing in the midst of it all, it was hard to imagine that it was more than 100 miles to the nearest village. Here in the wilderness, King Lacroix had built his own village. Families, mostly French-speaking, lived in white frame houses. Single men lived in the long beehive-like boarding house. A general store and the storehouse where supplies were divided for the various camps stood on the opposite side of the river.

A sawmill whined through the day to provide dimension lumber for the far-flung woods operations. A blacksmith shop glowed with the light of the forge as hammers banged and clanged all day to repair and build machinery, sleds, logging chains, and anything else that might be needed.

The muffled roar of a Lombard could be heard from inside a huge building as mechanics worked on its gas engine. The woods around the Depot hummed with the roar of tractors working in the forest, while heavy trucks rumbled over the wooden bridge that was part of the dam.

As it left Churchill Lake, the river added its own sounds to the din. The water splashing over the dam and the rapids roaring below were constant and unchanging. The dynamo at the dam that supplied the electricity to the settlement created a distinctive throbbing sound, which remained unnoticed except when it happened to stop. Then the silence could wake you from a sound sleep.

Jim, Mac, and the rest of the Spider Lake crew were staying at the huge boarding house until work on the river drive began. Key men were already being assigned and as expected, Jim was assigned to a bateau. His skill with these large river boats was well known on the river. Mac was made assistant on another bateau. His skill was nearly equal to Jim's, but the bosses were reluctant to put the lives of other men in the hands of someone who so willingly risked his own just for the thrill of it.

The bateaux were vital to the river drive. Unlike driving on a small stream, on the river the men worked from boats instead of from shore. Five- or six-man bateau crews worked the river from side to side and watched potential trouble spots. When a jam developed, the men got out and picked it while the bateau man stood by ready to take the other men off when the jam hauled. The men's lives depended on the speed and skill with which the bateau man could handle his craft.

M. T. Murphy came to the depot while the crew was there. He went out of his way to greet each of his men, make sure that all had been paid for their winter's work and that all were satisfied. M. T. sought out Jim and asked him to return to work early in the fall. Jim was pleased that M. T. asked him to take on the job of walking boss at one of his camps, and he quickly accepted. Already he looked forward to the coming season and his new responsibilities, but for now he was a river driver, and he was ready to get going.

The day before the Allagash River drive was to start, Jim and some of the other bateau men took the two cook's bateau and the wangan bateau down through Chase's Rapids. Most of the time the cooks and wangan man handled their own boats, but they were not bateau men, and no one wanted one of these important boats or their contents lost in the rapids. The booming crew went along so it would be ready to receive the logs when they reached Umsaskis Lake. There the logs would have to be boomed and towed to Long Lake Dam.

Almost two hours before daylight the following morning, the gates of the dam were opened so there would be plenty of water to float the logs to Umsaskis. At dawn the order was shouted, "Break 'em in." The huge piles of long logs along the river were rolled into the water, and sluicing began through the dam. The bateau drivers moved down river to their assigned areas, and the Allagash drive was under way.

Jim's crew was assigned to the area just above the big eddy. This bony stretch of water was often troublesome, especially if the water dropped a little. As long as the logs ran well, all the crew had to do was hold the 35-foot boat in a quiet eddy near shore and watch the wood float by.

Jim stood in the bateau, his pole braced to hold it in place, and thought about these strange men called river drivers who sat talking among themselves.

Maine State Library collection.

Bateaux were used to transport men, supplies, and equipment that had to be moved daily during the drive.

While he knew he was one of them, he could think about them as if he were an outsider. How natural they seemed here, he thought, the way they act, dress, and work. But how out of place they seem when they hit town. The men dressed near enough alike that they seemed to be in uniform. They wore heavy leather-caulked (they called them corked) boots. Their pants were cut off just below or above the knees so the pant leg wouldn't catch on their caulked boots. Usually the pants were held up by wide suspenders and the shirt was of faded flannel. Their long underwear provided the only leg covering below the knees. A hat was part of each man's costume and the only item of clothing that gave him some degree of individuality. Selections ranged from derbies to top hats.

Only the hardiest men toughed the drive from landing to boom. They were in and out of the cold water day in and day out, working in the rain or late snowstorms, and almost every night they crawled to bed soaking wet. Most of the men who stuck it out spent a few wild weeks in town and then, broke, headed back for another season in the woods.

Jim knew he was one of them. He fit in well, but he wanted other things too. He had saved most of his earnings over the years. Not for something special, but just because he couldn't see blowing it all and having nothing to show for it. Maybe someday he'd have his own operation, but these days it took a lot of money to buy the Lombards and tractors and other equipment needed to run even a medium-sized job.

Maine State Library collection.

Drivers pulling jill-pokes out of the willywags.

Jim's crew moved up and down their assigned area, keeping the logs moving. Whenever possible they tried to straighten out a jill-poke before it caused other logs to pile up on it. A jill-poke was a log caught crossways or otherwise out of the mainstream of moving logs. A jill-poke in the willywags was a log out of the mainstream—a log out of its element. The drivers also referred to an inept person as a jill-poke. And if he seemed really odd and out of place, they would call him a jill-poke in the willywags.

Over the roar of the water the crew could hear a man singing at the top of his lungs. Looking upstream they saw old Arnold Baker riding a log, singing with his black pipe clamped between his teeth. They didn't know why he was out there. Probably he'd been picking a jam and couldn't get back to his bateau. Everyone knew Arnold, and his skill on a floating log was legend. He was enjoying his ride now, and when he saw Jim's crew watching, he started clowning for their benefit. Through pantomime he indicated he didn't know

Maine State Library collection.

Old-time river driver. Still smiling in spite of the harsh conditions in which he lived and worked.

A driver riding a log through a driving dam's sluiceway. Their skill at this balancing act was legendary.

where his bateau was. He hopped into the air, did a twist, and landed on the log facing upstream, singing all the while. Then the log, following the current, moved closer to shore and beneath an overhanging tree. Arnold hit the tree and went sprawling into the water. "The door closed"—he went under. When he came up the black pipe was still clenched between his teeth, and he was singing louder than ever. Jim moved the bateau to pick him up and put ashore while he and the crew roared with laughter.

As the days went by, the crews kept the logs running and tried to anticipate problems. They pulled logs from poke logans, and moved logs that were whirling in a roozling hole out into the mainstream. Whenever they spotted a jill-poke they tried to get to it as quickly as possible. The logs that had become jill-pokes in the willywags were left for the rearing crew to roll in.

When the rear started moving down Chase most of the bateau crews moved down to Long Lake Dam, and soon the sluicing began there. This seven-hundred-foot earthen dam had been built in 1907 by the St. John Lumber Company. The company had also dynamited many rocks and ledges in the river, but plenty remained.

Just above Long Lake Dam was not Long Lake, but Harvey Pond, named after John Harvey, who had cleared a farm on the western shore. Some of the older men had known old John, and Jim always chuckled when he heard them describe him. "Came upriver in 1880 wi' axe n' oxen. Thin as a splinter n' ugly 'ern a meataxe. Married a old squaw n' had a dozen er more kids. Ast if tha kids were all his'n he said, 'speck so, leastways they was all cought in my trap."

The drive continued down river, the long days and short nights blending into a blur in Jim's mind. There were no serious incidents, so far, and in a matter of a few weeks, Jim wasn't sure just how long, the rear was through Round Pond, and the big trouble spot lay ahead: Allagash Falls. Meanwhile, the routine continued.

The meals were the one bright spot, or rather four bright spots, in the river driver's day. He might go to bed at night bone-tired and soaking wet, but he never went hungry. Keeping him fed was an operation in itself.

The cook, his two cookees, the clerk, the drive boss, and the wangan man traveled together in three or four bateaux. The larger the drive, the larger the wangan, as the whole entourage was called. Some even included a blacksmith bateau with spare peavies and other tools, as well as the equipment to repair them. The crew was housed in tents, usually large wall tents. The cook and cookees shared a small tent, and the boss and clerk shared another. The wangan consisted of these tents, rolls of bedding, clerks' records, some of the men's belongings, and variety of housekeeping paraphernalia.

The cookees set up the cook's tent, then built a fireplace with two rows of stones that served as andirons and a bar, supported by two crotched sticks driven into the ground, on which the kettles were hung on iron hooks. The cook began to roll out sourdough biscuits and gingerbread on his large bread board on top of a flour barrel. These baked in big reflector bakers on each side of the open fire. The cookees cut wood, dug a hole, and filled it with wood.

Maine State Library collection.

A river driver with the tool of his trade—a peavey.

Later they would place the big kettle of beans in the hole so the beans could bake overnight.

Meanwhile the wangan man pitched the sleeping tents and made up two beds in each, with an upper and lower spread running the length of the tent. With wood for a fire in front of each tent, the wangan was ready for the night.

The drivers worked from daybreak to sunset, seven days a week. Long before daylight, sourdough biscuits were baking, the steaming hot kettle of beans was pulled out of the ground and, at the cookee's cry of "Turn out!" the men stumbled from their bunks. After a hearty breakfast of beans, bread, molasses, gingerbread, and tea they trooped off to their stations with their peavies on their shoulders.

When the dishes were out of the way, the cookees packed the first lunch, which they would take to the men at about 10 o'clock. This meal, identical to breakfast, was carried in the wooden buckets and tin pails in which the beans and tea would be heated on the riverbank. The clanking of the tin cups strung together over his shoulder always announced the cookee's arrival. Once the first lunch was delivered, the cookees hurried back to the wangan to pack the second lunch, which they would serve in the same way at about two o'clock.

At dark the men returned to the wangan for a supper of salt codfish, salt beef or thick pea soup, with the invariable biscuits, molasses, gingerbread and tea.

Jim recalled the long hours he had put in when he worked as cookee of a drive. The day was full when the wangan was stationary, but when it moved many of the chores had to be postponed and the cookees worked far into the night to prepare the breakfast meal. The cookee's job separated the men from the boys in a hurry, for sure!

Drivers used booms to keep the logs on course. Booms were a string of long logs chained together. The ends of these floating chains of logs were attached or floating free depending on their purpose. There were several types of booms, each named according to its function. The general term applied to all types of booms, but it usually referred to the line strung across the river from pier to pier that would prevent the logs from floating farther downstream once they reached their destination.

"Bag boom" referred to the full circle of chained logs that enclosed floating timber on a lake, so named because of the bag-like shape it formed as it was towed.

To prevent logs from piling up in a cove or a backwater, a "sheer boom" was used. Both ends of this type of boom were hooked to the riverbank or to piers.

Another boom much like the sheer boom was the "fin boom." It differed from the others in that only one end was attached to shore. The free end was held in place by the river's current. On the pulp drive running down the West Branch, drivers used a "bathurst boom," made of squared timbers piled one on top of another to keep the little pulpwood logs from slipping under.

At Allagash Falls a sheer boom extended all the way across the north channel. This was no ordinary sheer boom. It was fortified with planks and strong enough to withstand impact from logs striking it in the heavy current. The boom was connected to a pier in the center of the falls. On the south side of the falls a fin boom kept the logs from getting tangled in the ledges at the head of the falls. On the ledges right above the falls the men worked on a pier or platform.

In spite of the preparations, the falls area easily became a bottleneck and all drivers dreaded the work here. Jim and his crew were a mile or two upriver from the falls when they got word to come down and help clear a jam. When they rounded the last bend before the approach rapids they could see the jam on the right side. Logs were still running through the channel but, unless the jam was released, it would be just a matter of time before the channel blocked up completely.

The problem with picking a jam at the falls was that when it hauled, the men had no place to escape. The falls dropped forty feet. Below were rocks, not water—not that a man would have a chance to survive in either case.

Maine State Library collection.

Log jam at Allagash Falls.

Maine State Library collection.

Maine State Library collection.

Working the drive. The drivers had to move the logs, which were hung up high and dry, back into the water.

Each jam was handled individually, but two techniques were commonly used. A man could climb out to the jam and choose what he thought might be the key log. Then he would tie a heavy rope to it, get out of the way, and let the men on shore pull the log free.

Another method was to lower a man on a rope to the jam and let him try to pick it free with his peavy. When the jam hauled, the men on shore jerked the volunteer into air off the moving logs and onto the shore.

When Jim arrived they were lowering a man over the jam. He didn't need to look to know it was Mac McGargle. And Mac was enjoying every minute of it! The rope on which Mac was suspended ran through a pully on a pole that sat at about a 45-degree angle from the top of the cliff on the south side of the falls. By controling the angle of this pole the men could put him down exactly where he was needed and, when necessary, the men on the other end of the rope could back up and pull him clear.

Mac landed, light-footed as usual, on one of the logs and went to work. He loosened several logs and with his usual agility leaped out of the way as they tipped and fell to the water and rocks below. Then he hooked his peavy into a log that barely moved when he pushed against it. But he could feel an ominous movement in the whole mass. The movement was so small that only Mac, standing on the logs, could detect it. Mac knew he had the key log, but he just had to make the most of the dramatic moment. He moved the key log a little more, and again the pile shifted imperceptively. Again! This time the grinding movement gained momentum. Mac scrambled to the top of the pile and stood there, arms in the air, while the mass of logs moved under him. He had already performed a dance or two on the moving logs before the men on shore realized what was going on and jerked him into the air, free of the hauling jam. Mac laughed and yahooed all the way to the shore. The drive boss gave him the usual tongue-lashing, but as always, the words rolled off him like water from a duck's back. Jim walked up to his friend shaking his head. Mac just grinned.

A channel had been blasted in several places below Allagash Falls to make it easier to drive the logs through this bony stretch of river. Still, the days that followed brought one jam after another. Jim and his crew were bone-tired from the long days of rolling, dragging, and lifting the heavy logs to keep them moving. Progress was slow, and the rear had not yet cleared the falls when word came upriver that there was a bad jam about four miles downstream from the falls.

Several bateau crews were already picking at the jam when Jim and his men arrived. Jim maneuvered his bateau below the jam, let his crew out, and took the boat ashore. The jam was on a ledge that ran to shore, so there was no need for more bateaux. Most of the men would run for shore when the jam hauled.

Other bateau crews remained in the water above the jam, pushing the logs out into the current so they would bypass the jam.

In the middle of the jam pickers, as usual, was Mac McGargle. He worked harder than anyone else, anticipating the excitement of the jam haul. The men picked at the stranded logs for nearly two hours. The boss was ready to send for dynamite when the jam started to move. After the first shudder of the logs everything moved fast—much faster than usual. The skillful bateau men quickly retrieved the men on the river side of the pile-up and then let the current sweep them out of the way. The men on the shore side ran across the moving logs and heaved a sigh of relief when their caulked boots landed on solid ledge.

When they turned back to watch the jam haul, no one was surprised to see Mac McGargle out there performing his usual dance with death. But this time things were moving fast, even for Mac, and he was having a hard time keeping up with the twisting, heaving logs. Just when he thought he had found a stable, level log, one of the logs from the tangle above landed on the back end of the log Mac was standing on. Mac was catapulted upstream into the moving mass of logs. The other drivers watched the grim scene, with their mouths hanging open, helpless. They knew better, but they almost expected to see the agile McGargle land on his feet on the top of the pile and continue his dance. But, he didn't. Mac fell into the churning mass. His body was pulled from the water an hour or so later about half a mile downriver.

It almost seemed to Jim that Mac would have wanted to die that way. He would miss his friend. Mac cheered him up when he was feeling low, because he never seemed to have a care in the world. Jim made a crude cross and placed it over his grave and hung Mac's heavy leather boots from it. Jim had seen men killed on drives before, but Mac was the first close friend he had lost. It hit him hard.

The rest of the drive down the Allagash and St. John was one of steadily increasing problems as the water slowly went down. In his grief, Jim worked like a demon, as if he alone could hurry the drive along. All he could think about was getting to Fort Kent and off this cussed river. He knew he'd be back, but for now he wanted out! Finally, after almost six weeks on the river, the rear was brought to the boom, and Jim drove his bateau onto shore for the last time this season.

When Jim arrived in Fort Kent, where his folks lived, many of the drivers were already in town celebrating. He smiled as some of the men he had worked with for months walked by without even noticing him, a girl on each arm and a bottle in each hand. They would be tolerated here only as long as their money lasted. Then, back to the woods. Here, these men were jill-pokes in the willywags.

When The Allegash Drive Goes Through
by Holman F. Day

We're spurred with the spikes in our soles;
There is water a-swash in our boots;
Our hands are hard-calloused by peavies and poles,
And we're drenched with the spume of the chutes.
We gather our herds at the head
Where axes have toppled them loose,
And down from the hills where the rivers are fed
We harry the hemlock and spruce.

We hurroop them with the peavies from their sullen beds of snow;
With the pickpole for a goadstick, down the brimming streams we go;
They are hitching, they are halting, and they lurk and hide and dodge,
They sneak for skulking eddies, they bunt the bank and lodge.
And we almost can imagine that they hear the yell of saws
And the grunting of the grinders of the papermills because
They loiter in the shallows and they cob-pile at the falls,
And they buck like ugly cattle where the broad deadwater crawls.
But we wallow in and welt 'em with the water to our waist,
For the driving pitch is dropping and the Drouth is gasping "Haste!"
Here a dam and there a jam, that is grabbed by grinning rocks,
Gnawed by teeth of the ravening ledge that slavers at our flocks;
Twenty a month for daring Death; for fighting from dawn to dark—
Twenty and grub and a place to sleep in God's great public park;
We roofless go, with the cook's bateau to follow our hungry crew—
A billion of spruce and hell turned loose when the Allegash drive goes through.

My lad with the spurs at his heel
Has cattle-ranch bronco to bust;
A thousand of Texans to wheedle and wheel
To market through smother and dust.
But I with the peavy and pole
Am driving the herds of the pine,
Grant to my brother what suits his soul,
But no bellowing brutes in mine.

He would wince to wade and wallow—and I hate a horse or steer!
But we stand the kings of herders—he for There and I for Here.
Though he rides with Death behind him when he rounds the wild stampede,
I will chop the jamming king-log and I'll match him, deed for deed.

And for me the greenwood savor and the lash across my face
Of the spitting spume that belches from the back-wash of the race;
The glory of the tumult where the tumbling torrent rolls
With a half a hundred drivers riding through with lunging poles.
Here's huzza for reckless chances! Here's hurrah for those who ride
Through the jaws of boiling sluices, yeasty white from side to side!
Our brawny fists are calloused and we're mostly holes and hair,
But if grit were golden bullion we'd have coin to spend, and spare!

Here some rips and there the lips of a whirlpool's bellowing mouth,
Death we clinch and Time we fight, for behind us gasps the Drouth.
Twenty a month, bateau for a home, and only a peep at a town,
For our money is gone in a brace of nights after the drive is down;
But with peavies and poles and carefree souls our ragged and roofless crew
Swarms gayly along with whoop and song when the Allegash drive goes through.

["Allegash" is one of several different spellings that have been used for what we now know as the Allagash. The word is an English translation of the Wabnaki word for "Bark-Cabin-Lake." As a result, every author came up with his own spelling upon hearing the word. Thoreau, writing of his 1857 trip on the Allagash and West Branch, also used the "Allegash" spelling.]

[CHAPTER THIRTEEN]

THE EAGLE LAKE & WEST BRANCH RAILROAD

The first two decades of the twentieth century saw the beginning of a great change in the cutting and moving of wood in the Allagash. In 1912 the annual cut of pulpwood outstripped sawlogs for the first time. The life of the woods worker or the riverman did not really change all that much, but it seemed to from their point of view. Many of them hated this small stuff, as the four-foot bolts were called, and many of them moved on to other forms of labor. No self-respecting logger was going to stoop to bear-hugging four-foot sticks of firewood out of the brush. The driver, accustomed to walking high and dry across the logs in rivers and streams, found himself crawling out of the water like a drowned rat after stepping, without thinking, from a long log to a stick of pulp. No amount of river driver's skill would keep the little bolt of wood afloat under his weight.

The Great Northern Paper Company played a prominent role in Allagash lumbering in the 1920s and '30s though the company owned no land here. Great Northern contracted with Eduard LaCroix, who did own land in the Allagash area to furnish pulpwood to its Millinocket mill. LaCroix and Great Northern faced the same old problem that had plagued lumbermen in the Allagash for the previous 80 years—how to get the wood from the Allagash watershed, Eagle and Churchill lakes, into the Penobscot watershed, which now included Chamberlain and Telos lakes due to the dams built earlier on those lakes. Great Northern's mill was located on the West Branch of the Penobscot just upstream from its juncture with the East Branch. Because of the mill's location and other problems connected with driving the East Branch, Great Northern concentrated on driving the West Branch, where pulpwood could be floated right to its mill.

A number of alternatives were considered. The company considered restoring the tramway, for instance, but decided against it because the tramway moved too slowly to provide the quantity of wood the company needed within the time it required it.

In 1921 Fred Noad surveyed the area around Umbazooksus and Chamberlain Lakes. He noted the low height of land between the two watersheds and came up with two plans. The first, and the one he favored, called for digging a canal to connect Mud Pond and Umbazooksus Lake. Eben Coe had thought of doing this in 1841, but he had been beaten to the punch when Telos Dam was built. Noad's alternate plan called for building a railroad between Eagle Lake and Umbazooksus Lake.

It was as Noad was drawing up his plans that Eduard Lacroix, a Canadian known as King Lacroix, came onto the Allagash scene. Lacroix, or Le Roi (The King) as his French woods workers called him, had gained his name and reputation in Ottawa where he was a popular politician. He carried the title into the Maine woods where woods workers, Canadians, and Mainers alike, held him in high esteem.

Lacroix once employed as many as 3,000 men in the north woods of Maine, many of them French Canadians. His honesty and fairness earned him a good reputation with everyone in the Allagash area. He believed in paying the men well, furnishing the best equipment, providing clean, comfortable living quarters, and feeding the lumbermen the way they liked to be fed: hearty meals. His clerks and scalers lived in houses he built for them, and his lumbermen lived in big boarding houses. The following tribute, author unknown, sums up how his workers felt about him.

The Man of the Hour

These lines are composed by a scaler, it seems,
Who is scaling the pulp wood at Thoroughfare Stream,
Where the young and the old and the low and the high
Are singing the praises of Eduard Lacroix.

He sure is some hustler to corral all these means
In order to purchase such costly machines;
His mountains of pulp are a wonderful sight,
And tractors are humming by day and by night.

In 1925 Lacroix contracted with Great Northern Paper Company to deliver 125,000 cords of pulp per year from the Eagle Lake-Churchill Lake region to West Branch waters. This contract brought about the most ambitious enterprise ever attempted in a remote area, miles from the nearest road or railhead. The enterprise was to become known eventually as the Eagle Lake and West Branch Railroad, or the E. L. & W. B. R. R. Great Northern and the Madawaska Company (Lacroix) were partners in the venture.

Early in 1925 Lacroix, F. A. Gilbert, A. V. MacNeill, and O. A. Harkness made a canoe trip down the East Branch to Grindstone to take one last look at that branch of the river before committing themselves to the West Branch. This group of able, practical men found the estimated cost of preparing the

East Branch for driving prohibitive. There were a number of large drops in the river, and several large dams would have to be built at strategic points, as well as abutments. This coupled with the advantage driving on the West Branch prompted them to abandon plans for the East Branch once and for all.

According to the original plans, the railroad was to extend from the tramway on Eagle Lake to the Great Northern terminal on Chesuncook, eliminating the need to drive Umbazooksus Lake and Stream. Lacroix was to build the 11 miles to Umbazooksus, Great Northern the remaining five miles to the terminal. However, the lower five miles were never used to transport wood, only to haul supplies. No one seems to know why this was so.

King Lacroix built three camps, at the tramway, at Ellis Brook, and at the head of Umbazooksus Lake. The men, supplies, and equipment for these camps came in by way of Lac Frontiere on the Maine-Quebec border some 50 miles away. Lacroix had built this road connecting the Canadian town with Churchill Depot.

The railroad right-of-way was cut late in the fall of 1925, and during that winter 15,000 cords of pulp wood Lacroix's crews had cut around Ellis Brook were hauled by tractor and unloaded on the ice of Umbazooksus Lake.

The most remarkable piece of construction of this whole railroad project was the trestle across Allagash Stream at the northern end of Chamberlain Lake. Ten feet above water, the trestle stretched 1,800 feet. Five piers with 20-foot spans supported the trestle. The rest of the structure was made up of 10-foot bents. C. M. Hilton engineered the trestle; Lacroix's crew built it. It was a masterpiece of backwoods engineering.

In the spring of 1926 Lacroix's crews started grading and laying steel. As no rolling stock had yet been brought into the woods, a tractor was fitted with wheels to fit the tracks. It was used to haul rails and other materials needed for construction.

Terry Harper collection.

Part of Locomotive No. 2 on its way to tramway. The locomotives traveled behind gas Lombards over LaCroix's all-season road, which is still in use today. The entire journey from LacFrontier, Quebec, took three days.

Work on the railroad continued into the winter. The track was laid by freeze-up, but the 600-foot trestle extending the end of the line out into Umbazooksus remained to be built. Construction started that winter and was completed by the following June.

During the winter of 1926, while the work on the railway was underway at Umbazooksus, an almost unbelievable feat of moving heavy equipment was taking place between Lac Frontiere and the tramway. The hard frozen ground and smooth frozen surface of the lakes were of paramount importance to this gigantic task. Steam and gas-powered Lombard Log Haulers were teamed up to haul the two steam locomotives, one weighing 100 tons, the other 90 tons. They also hauled two 18-ton gasoline-powered Plymouth switchers and the 60 flatcars which would eventually be converted to haul the pulpwood.

The heavy equipment rode on specially built sleds. The power required and the problems encountered must have been awesome. But there is no record to indicate that the men involved in the project considered it overwhelming. These men were used to hard work. They probably took this massive undertaking in stride.

At the tramway camp crews had plenty of work. They erected two conveyors that would lift the pulpwood from the water to the pulp cars.

Walker Transportation Museum.

The Eagle Lake and West Branch Railroad in operation.

Another huge project was remodeling the flat cars and operating the saw mill that furnished the necessary lumber for the entire operation. The steam locomotives were converted from coal to oil during this period because it reduced potential fire hazard from cinders. Oil was also easier to transport than coal and so permitted rapid servicing during round-the-clock operation.

Remodeling of the flat cars consisted of modifying the floors so they tilted 12 inches to one side. High rack bodies were built on the cars and hinged at the top on the low side of the tilt. So when the car traveled the 600-foot trestle at Umbazooksus, where one track was six inches lower than the other and the side of the rack body was released, it created an 18-inch tilt, allowing about a quarter or a third of the wood to drop into Umbazooksus. This made it much easier to unload the cars.

The railroad, which was owned by Lacroix when it was completed, was named the Umbazooksus and Eagle Lake Railroad, or U. & E. L. R. R. Before it went into full operation Great Northern bought the railroad and all its equipment and promptly changed the name to Eagle Lake and West Branch Railroad, or E. L. & W. B. R. R. Great Northern retained Lacroix to manage the entire operation.

The first load of pulpwood moved across the divide on August 1, 1927. By the end of the first week in operation, 1,800 cords of wood were dropped into Umbazooksus Lake. At this rate, it became apparent that the operation could handle only a fraction of the 125,000 cords of pulp that had to be moved each year. Efficiency had to be increased.

The train required at least three hours to make its run, often longer, because of the numerous curves in the road. No time could be cut here.

Terry Harper collection.

Moving pulp onto one of the conveyors on Eagle Lake.

Walker Transportation Museum.

The two ends of the E.L & W.B.R.R.. Above is the loading of the pulp cars. There were three conveyers, one spare. Below is the wood dump at Umbazooksus Lake. The lean of the tracks and of the pulp car can be clearly seen in this photo.

Walker Transportation Museum.

To increase handling speed, railroad officials replaced two 10-car trains with three 12-car trains, as was called for in the original plan. The three trains made it possible for one to load while the other two were hooked up to the engines coming and going. Original plans also called for hauling round the clock, and this schedule was implemented when the installation of an electric lighting plant made night loading feasible.

Tim & Sue Caverly collection.

The engines were housed in this shed when they were cooled down for the last time. The shed was burned down and the locomotives have stood in the elements ever since.

Storage towers for water and oil provided another time-saver. These towers made it possible for the engines to be supplied with both water and oil in less than 10 minutes.

These time-saving measures combined to improve efficiency by nearly 400 percent, so an estimated 6,500 cords could be hauled in a single week.

The day-to-day operation of the railroad from water to water went like this: The men fed the pulpwood floating in Eagle Lake with pickpoles into two conveyors. The 225-foot conveyors raised the wood 25 feet. Each conveyor was capable of delivering a cord of wood from water to car in one and a half minutes, which meant that it took 18 minutes to load a car to its capacity of 12 ½ cords. A third conveyor stood by as backup. The wood was not stacked in the cars, but rather allowed to fill as it dropped from the conveyor. The cars could have carried much more wood if the logs had been tiered, but the extra time it would have required didn't make stacking feasible.

One of the Plymouth engines was used to do the yard shifting, so when a steam engine returned with its load of empty cars, the gasoline engine moved them to the free track for loading. After fueling, the steam engine went on the road again. At the turn-out track about halfway along the route, the empty train could pull off and allow the full one to pass.

At the Umbazooksus terminal the loaded cars were backed onto the tilted trestle. The steam engine with its train of empty cars returned to the tramway. Another Plymouth engine did the shifting at this end.

The wood that did not fall into Umbazooksus when the sides of the cars were released was pushed and pulled out by men using hooks and poles. It took about an hour to unload the 12 cars, which, when empty, were pulled out to the main line by the shifter.

The bark that accumulated in the water beside the trestle created a problem. So much bark gathered that it hindered the flow of pulpwood. So a scraper was designed to push the bark ahead of the pulpwood.

The August 1928 issue of *The Northern*, a Great Northern Paper Company publication, carried the following item.

"Eagle Lake Tramway" A. G. Faulkner, Correspondent

On June 12 we commenced hauling wood. The first wood moved was a small lot of about 2,100 cords cut on Allagash Stream last winter and loaded at Allagash Stream trestle.

Hauling on Eagle Lake began on June 20. Since that date this place has been a scene of activity, there being two full crews employed, one by day and one by night. Unless some minor delay occurs work never ceases from midnight Sunday until midnight the following Saturday. Mr. August Lessard is superintendent in charge and Mr. Florian Poulin is the head clerk.

Two oil burning steam locomotives are employed in the hauling; the new one weighs about 100 tons and works to perfection. The other weighs ten or fifteen tons less and is used as a spare. In addition to these, there are two gasoline locomotives of eighteen tons each; one is used in the Tramway yard for shifting cars and the other is used for the same purpose on the landing at Umbazooksus Lake. These machines are smart with plenty of power and make ideal equipment for yard work.

Under favorable conditions, such as fair winds for towing booms in from Churchill Lake to the conveyors and also for unloading the cars at Umbazooksus Lake, we can haul between eight and nine thousand cords per week.

About one hundred men are employed here including the men who work towing in the booms from Churchill Lake. Our little colony consists at present of the following permanent residents: Miss Annett Poulin, Mr. and Mrs. Mathias Morresette and children, and Mr. And Mrs. William Dubay of Old Town, Maine. Mr. Dubay is employed by the Maine Forestry Department as chief fire warden in this section. We also have, of course, a large floating population of men coming and going as in every operation.

A casual inspection of our several well-kept gardens would surely impress a visitor with the thrift and industry of our inhabitants. All the credit in this connection is due to the ladies, who judging from their planting designs and present healthy conditions of the crops, are natural born farmers.

We have as regular visitors Mr. Charles LaCroix of St. George, P. Q., Mr. E. W. Vickery of Greenville and Mr. George Gruhn of Augusta who is "commander in Chief" of the fire fighting forces in this part of the state.

During the past two weeks Mrs. Bert Duty and daughter, Ruth, of Greenville, have been visiting Mr. Duty, a state game warden, who has headquarters at Eagle Lake near the Tramway. The ladies returned to their homes on Saturday, July 14.

The third conveyor and the extra cars for each train must have been in use by the time Mr. Faulkner wrote his report for *The Northern*, because he estimates the weekly total of wood hauled at eight to 9,000 cords—a considerable increase over the previous high of 6,500 cords. He referred to the smaller engine as a spare, but it must have been in constant use for the operation to deliver up to 9,000 cords to the Umbazooksus terminal. Even without down time (highly unlikely) one locomotive could not have accomplished this much.

Families at tramway were housed in frame houses provided by King Lacroix. The single men lived in a large boarding house. Some of the old cabins from the tram era were still in use. It was a lively, busy little community whose sole purpose was to move wood over the height of land between the watersheds. The closest neighbor was Churchill Depot, a similar community, whose job it was to cut the wood that had to be transported. If this region was a wilderness, no one who lived there then noticed.

The Eagle Lake and West Branch Railroad operated until 1933 when the two locomotives were retired to the large house at tramway. The boilers were blown down and cooled off forever. This railroad, the most incredible wood-moving scheme in the Allagash, was abandoned to the forest.

An article in the November, 1927, issue of *The Northern* stated, "Inasmuch as conservative estimates allow at least 20 years to be required for cutting of pulp to be marketed in this way, it will be seen that the usefulness of the railroad in the wilderness has only begun."

It is unbelievable that men who could conceive, build, and operate such a fantastic venture as this wilderness railroad could have been 14 years off the mark in their estimate how long it would take to do the cutting. Perhaps the writer added that final paragraph to his article to pacify nervous stockholders. That, and more efficient means of moving wood were coming along.

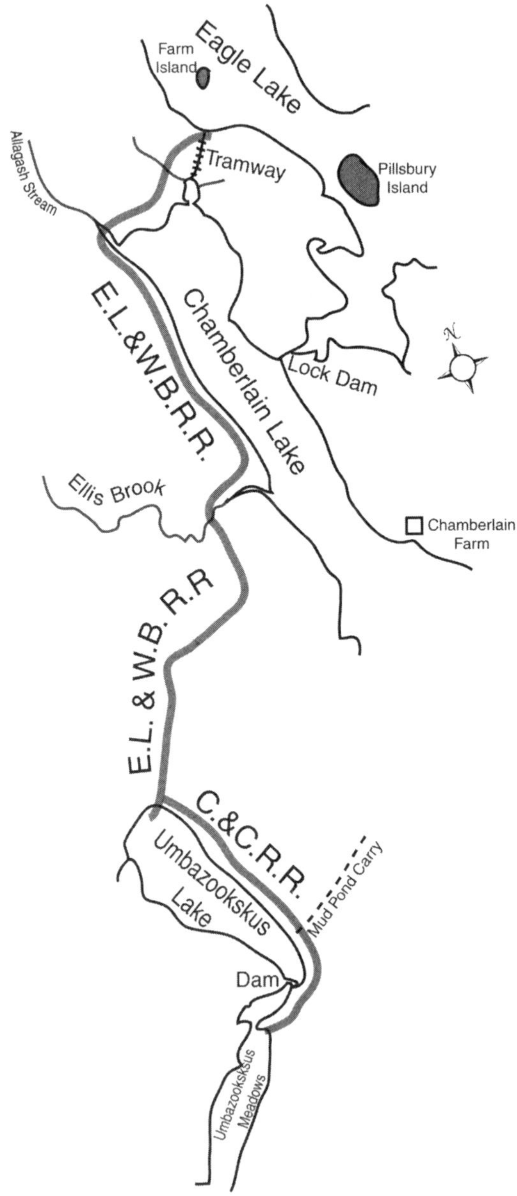

Routes of the E. L. & W. B. R. R. and the C. & C. R. R. The original plan was to bypass Umbazooksus and dump the wood directly into Chesuncook Lake, but that never happened.

[CHAPTER FOURTEEN]

BIG JIM'S CREW

The sun was not yet showing above the eastern horizon when the 65-foot horseboat started to climb Casey Rapids, the second loop of a big S the Allagash River makes before it enters the mighty St. John at the little village of Allagash. It was mid-July, but recent rains gave the river a good pitch of water, and the fresh horses pulled the big boat along easily with the help of the two pike men on board.

Standing in the stern leaning on the sweep was the owner of the boat, Big Jim Fleury. Big Jim was 42 years old and had been involved in hauling freight by one means or another since he was just a lad helping his father tote supplies to lumber camps up and down the St. John. At 15 he landed his first job as pike man on a horseboat, and from that time on he knew there would be no other life for him. Now he owned three boats and had contracts to supply a half dozen upriver lumber camps. His biggest contract, and his destination on this trip, was Churchill Depot.

Jim was a big man, 240 pounds. His fists were as big as hams, and he didn't mind using them when the occasion warranted it. Many a wood cutter heading upriver on Jim's boat found himself stranded in the river because he had refused to help the crew in a bad spot. A hard man Jim was, but fair. That's how his crew received him, and they had been with him long enough to know.

Two of Big Jim's crew were Tim Pelkey and Henry Bradley, both pike men. Their job was to help the horses move the big boat upstream with their long poles and also help with steering when necessary. Both Henry and Tim had worked horseboats on the Allagash so long that they could help maneuver the boat through the tightest spots while they carried on a conversation about some floozie in a Fort Kent saloon.

Augie Frye was the halyard man. His job was to tend the long line that went from the boat to the horses. He also helped with a pike pole when necessary as did everyone else on board, including Big Jim and even Ma Tukey in extreme cases.

Ma Tukey was the only woman horseboater on the whole Allagash. Ma did the cooking for the horseboat crew, and she did it so well that no one, not even Big Jim, interfered. Ma wasn't exactly the dainty type. She weighed about 150 or 160 pounds and always wore men's clothes. Her heavy wool britches were stagged between ankle and knee like the rest of the crew's and held up with wide, rugged suspenders. Her shirt was red and black checkered wool, pretty ragged, and dusted with flour most of the time. When Ma got ready to work in dough she dipped fir branches in fly dope and tucked them under her straw hat to keep skeeters and minges away.

Ma could cook. Oh, could she cook! Every man on the crew went out of his way to be nice to Ma because the better her mood, the better her cooking and the more generous she was with it. Ma made no secret of the fact she had a favorite among the crew. He was Alec Tidd the teamster, a horse-tender, or "horsefly" as these men were called on the river boats.

"Little Alec," Ma called him, and the name fit. The wiry little teamster weighed only about 120 pounds, but what he lacked in size he made up in energy and good nature. His appetite for Ma's cooking belied his size, and he kind of liked her fussing over him and "motherin" him, so to speak.

Alec was a sight to see out there with his horses, agile as a circus bareback rider. Horses were the love of Alec's life, and they in turn responded to him lovingly. Alec never raised his voice to a horse except in song, and when he spoke the horse responded instantly. Big Jim knew he had a good man in Alec, and he knew too that he could count on him as long as he kept him supplied with good horses.

Big Jim's boat made good time on this trip, and by mid-morning it was pulling by Ghost Landing Bar. Near here, in the flourishing pine days, a logger was killed while felling a big tree. The tree was hauled to the river, but when it was discovered that its core was rotten, it was left on the shore. From that time on river drivers swore whenever they passed the site they saw the ghost of the

Terry Harper collection.

A horseboat making its way up the Allagash. The "horsefly" can be seen on the back of one of the horses in the upper left of the photo. The canoes on board probably belong to some "sports" on their way up for hunting or fishing.

chopper standing on the river bank, beseeching them to turn the log into the river so his soul could rest.

Two or three miles above Ghost Landing Bar the boat reached McGargle Rocks where the crew tied up for dinner. At this rate they would reach Allagash Falls long before dark, and with luck they could have the portaging done today instead of in the morning, as was the usual routine. The boat was no sooner tied up in the quiet water below the rocks than Tim and Henry were ashore, scurrying to get the cooking fire going.

"Set me a green pole to hang this tea water on," Ma ordered, "and git some rocks to put this spider over the fire." The men, always ready to please Ma, obliged quickly.

Ma had everything ready to go on the fire before they got off the boat, and in less than an hour after they tied up the crew polished off a meal of potatoes, canned beef, bread, and applesauce. While the others answered Ma's swift commands, Alec Tidd gave his charges their noon ration of oats. While they ate he rubbed them down, talking to them all the while. Lucky were the horses under the care of Alec Tidd.

"Come an git it now, or you'll git it tomorra," Ma yelled as the men lined up to have their plates filled. When she had finished serving Ma picked up two plates, filled them, and called Alec to leave the horses and come to eat.

No one talked during the meal because the same rule that applied in logging camps throughout the north woods applied here: "No jabberin', no roostin', while hoggin." And Big Jim enforced the rule to the letter. The boat was soon underway again. Ma would wash dishes and then start supper. By the time they reached Allagash Falls she would have potatoes and vegetables peeled, and bread all mixed and raised, ready to bake in the large reflector ovens she carried. Sometimes, if she felt like it, and if the men brought in plenty of firewood, she would treat them to a fresh-baked cake after supper.

At McKeen Brook the boat passed the first of many lumber camps they would see on their way upriver. The river here was rather steep, and about half way up the pitch Alec yelled back, "Kee-rister ahead!" This meant that a boulder lay in their path. While Alec turned the team to avoid the rock, the crew leaped to the rail and shoved the boat hard to the same side Alec had traveled. When they had passed the menacing rock the sure-footed horses kept pulling and the boat crew kept poling.

The horses were large animals, about 1,800 pounds each, shod in soft iron shoes. Though they wore out quickly, the shoes stuck to the wet rocks and kept the horses from slipping and injuring themselves. At times when the water was deep and there was no place for the horses to go ashore, the men would get on the line, take up slack and then hold the boat fast with their poles while the horses swam through the deep stretch. As they swam the slack line was let out. When the horses had their feet on bottom again they resumed pulling.

Good horseflies like Alec knew every hole and large rock in the river, and they did everything they could to make things as easy as possible for the hard-working animals.

At Allagash Falls Jim kept a wagon to haul the cargo to the top of the carry. Alec had the horses hooked to it as soon as they were ashore, and by the time the big boat was tied up he was ready to start loading. The boat was carrying about six tons of goods for Churchill Depot. This was an average load for a boat this size. When the water was higher they could haul as much as eight tons, and when the level was at its summer low they would cut back to four tons.

Big Jim's loading list for this trip looked like this:

39 bales hay	6,000 pounds
26 bags oats	2,000 pounds
5 barrels flour	1,000 pounds
4 barrels sugar	1,248 pounds
2 barrels beef	500 pounds
1 barrel molasses	700 pounds
5 boxes clothing etc	500 pounds
4 bundles pick poles	200 pounds

The crew loaded the supplies and hauled them to the put-in point at the head of the approach rapids to the falls. Moving the heavy barrels and bales was hard work, but the men were used to it. They had worked together on many of these carries, so not a motion was wasted. Here at the falls, they would set up camp. Ma was already busy around her fire.

By the time the men carried the last load, supper was almost ready, so Big Jim decided to call it a day. "Take care a yer hosses, Alec," he said, "looks like Ma's 'bout ready fer us."

"OK," replied Alec, and he led the team to its accustomed spot at this familiar camp site. As soon as he pulled off their bridles he gave them their evening quota of oats, and then he unharnessed them. Before Alec would eat himself the horses would be fed oats, given hay, watered, rubbed dry, and provided with a canvas tarp in case of rain, a luxury not enjoyed by all Allagash tow horses.

When the men had eaten their fill of stew and fresh-baked bread, they sat around the fire and relaxed for the first time since early morning. This was their favorite time of the day, and as they sat around smoking their pipes they talked about whatever occured to them, keeping in mind that Ma was a woman no matter how she dressed.

"Wattaya 'spose they want 'ith all them pick poles up to Churchill?" Augie asked of no one in particular.

"Donno," replied Tim Pelkey, and as Big Jim approached their little circle he repeated the question.

"Them's goin to the old tram where King Lacroix has built hisself a reg'lar railroad to haul pulp wood over the height o' land to Umbazooksus," Jim answered. "Hear tell they's men workin round tha clock feedin them conveyors 'ith pulp to load tha railroad cars."

"Heard 'bout that railroad. Can't figure how they got all that heavy stuff inta tha woods," Henry said.

"They say it was all hauled over from Canady during the winter of '26," Big Jim said. "Used Lombard Log Haulers mostly; hear tell they got two steam engines that weigh 'bout a hundred tons each."

"If them big log haulers c'n haul loads like that around tha woods how long ya s'pose the hosses will be around?" said Augie as he grinned and jerked his head toward Alec.

Alec immediately rose for the bait and came to the defense of his beloved horses. "They'll never replace the woods hoss," he said. "How'r ya goin ta teach a tractor to step over a log?"

"They'll cut tha logs outa tha way," Augie replied.

" 'N I s'pose the tractor'll stop when a log fetches up?"

"They'll cut tha roads wide," Augie came back.

The good-natured goading continued until Alec realized he was being teased, and he grinned sheepishly and walked off to make a final check on the horses before turning in for the night.

Maine State Library collection.

A horseboat tied up below Allagash Falls. Sometimes two boats were used—one below the falls and one above. The goods were transported over the carry trail in a wagon. Thus, they avoided the job of moving the large boat up around the falls.

The following morning after breakfast they hauled the 65-foot boat up to the put-in point at the head of the rapids. The sturdy boat was moved overland on log rollers. On level ground the horses could pull it faster than the men could carry the log rollers forward to put them in front of the boat. To get the boat up over the steep part of the carry, they hooked it to a large steel ring set into the ledge at the top of the rise. By means of ropes and two sets of block and tackle the horses pulled the boat up by pulling downhill, one on each side of the boat. Because of the mechanical advantage of the six by double-two blocks the horses could pull the heavy load easily, although three different hitches were needed before the boat reached level ground.

Within two hours the boat was being loaded at the head of the falls. The cargo would be handled this way one more time, at Long Lake Dam, before reaching Churchill Depot, four days away.

Today they would pass several lumber camps before they tied up for the night. Just a couple of miles up from the falls was Michaud Farm, then came Cunliff Depot, then Five Finger Brook. These camps were more or less idle now, except for the depots, and would not be active until fall when the men returned for their winter's work in the woods. The depots kept their crews busy repairing machinery, storing supplies, and attending to other important details.

By supper time Big Jim's boat had reached Musquacook Deadwater. At this long, slow, deep stretch of water a path had been cleared on one shore and the horses pulled on dry land for a change. The pike men kept busy holding the large craft away from shore as they proceeded against the slow but persistent current. When they were almost to the point where Musquacook Stream entered the river, Jim called a halt and the boat was tied up for the night.

The crew quickly set up while Ma got supper ready. One large tent for the five men and a small one for Ma. Tomorrow was Sunday, their day to rest, so everyone was in a good mood. A bean hole was dug and later, after supper, Ma would put in a large kettle of beans to bake. The crew would be treated to fresh bean-hole beans on Sunday morning.

Baked beans have always been a traditional Saturday night supper for Mainers, but on a river boat there was little time for baking beans except when the boat laid over for Sunday.

While the crew ate, Ma kept the fire burning high with a good supply of hardwood that Augie and Henry had cut for her. When she had a sizeable heap of glowing coals, she shoveled a good layer of them into the hole. Next she put in the large iron kettle that held the beans that had been soaking all day along with her special fixings. She shoveled more coals into the hole around the kettle and over its cover. Finally she covered the kettle with dirt. The beans would bake through the night.

When work was done for the evening, Ma dug into the basket where she kept her "treasures" and fished out a harmonica. The crew sat quietly around the fire as she played songs like "Mocking Bird," "Skaler's Waltz," and the "Merry Widow Waltz." One by one the sleepy men stretched, yawned, and headed for the big sleeping tent. When they were all gone, Ma took a snort from the jug she always carried and headed for her own little tent, which was just big enough for her and her treasure basket.

Sunday was the day of rest, but there were a number of tasks done that day because there was no other time to do them. One of these chores was laundry. Along about noontime the makeshift clothesline was full and Ma's big bloomers, the only women's clothes she ever wore as far as anyone knew, were up there flapping in the breeze along with the men's shirts, pants, and long-handled underwear.

Sunday was also bath day. Ma always went off around a bend in the river to bathe while the men soaped up and skinny-dipped alongside the boat. After her bath Ma took out one of the flies she always kept in her hat and went fishing.

Ma always fried doughnuts on Sunday evening. The men stood around eagerly waiting for the fresh doughnuts to come out of the hot fat. Ma would let them have four apiece. If one of them dared to reach for a fifth, she would flick a little hot fat on him and he would run off howling. Somehow, though, she always seemed to be looking the other way when Alec reached for more than his quota.

By noon on Monday Big Jim's boat had just about reached Round Pond. As the boat passed Turk Island, just below the pond, Alec carefully steered the team around the deep hole that had been responsible for the island's name.

One night Joe Jalbert, riding a great white horse named Turk, fell into that hole. Joe swam to the boat, but his horse tangled in his harness and drowned. The place has been known as Turk Island ever since.

As soon as the boat reached the pond above the rapids the crew stopped for dinner. Then they took the horses aboard, dropped two weighted planks for leeboards, unfurled a sail, and sailed the ungainly looking vessel across Round Pond to where the river entered. Once again the boat was in its natural element of shallow, fast-moving water.

The big horseboat continued its slow uphill journey. Gradually the familiar landmarks slipped by; the lumber camps, the shallow rapids that always gave them trouble, the places that brought back individual memories for each crew member. Above Long Lake Dam they raised the sail again, and the heavy boat wallowed through the waves of Long Lake and Umsaskis Lake.

Each crew member worked steadily at his usual job. They knew they would be returning downriver for another load soon and that the whole process would be repeated. The downriver run meant a free ride for the horses, but not for the men. They worked as hard, or harder, on the northbound trip as they did going upriver. The boat had to be held under control and, with gravity and the current pulling on it, it required the joint effort of the entire crew.

The crew spent the last night out before reaching Chase's Carry at Chisolm Brook at the point where the river enters Umsaskis Marsh. The next day the boat would reach the beginning of the carry where the cargo would be put ashore. From there, teams of horses from Churchill Depot would provide transportation. Their supply wagons would bump over Chase's Carry and Big Jim would head down river with King Lacroix's credit voucher in his pocket.

That night the crew sat around the fire hoping Ma would play a few harmonica tunes, but she didn't. They noticed she had been acting kind of sober and withdrawn lately. "Little Alec," she said suddenly, "this is your's 'n my last trip."

Everyone's mouth dropped open. No one spoke. The only sound was the crackling of the fire. The crew just sat there and stared, openmouthed, first at Ma and then at Alec.

Alec gulped and said, "What makes ya think so, Ma?"

" 'Cause you 'n me is goin' ta git married. That's why, Little Alec Tidd," Ma answered as firm as you please. "I'm startin' up a restaurant in Fort Kent or Ashland. Maybe Bangor. Ain't made up my mind just where yet. But, when I do I'm goin' ta need a dishwasher. Now, you c'n still have yer hosses to play around with, but yer goin' with me."

All Little Alec could say was, "Huh."

"Yep," Ma said, "you'n me is goin' to git hitched right after this trip."

The rest of the crew shut up like a clam, but Big Jim groaned aloud. Here, in one fell swoop, he was losing not only the best damn cook on the river, but a horsefly without peer as well.

There was no harmonica playing that night. No more conversation either. Alec sneaked off to the tent and moaned and groaned all night. The rest of the crew quietly moved off to the tent, leaving Ma alone by the fire.

Tim Pelkey took one last look through the tent flap before crawling between his blankets. He saw Ma sitting there by the fire looking as pleased as punch, and he figured she took more than one swig from her jug tonight because it was sitting there on the ground beside her with the stopper out. Faintly he could hear Ma humming to herself, and he noticed her face looked almost pretty—well, at least happy.

The following is a questionnaire sent to Tom Pelletier of Allagash by Maine Guide Norm Wight in April 1966. Mr. Pelletier worked on Allagash horse boats for 38 years.

ALLAGASH RIVER BOAT:

Length? 65 feet, Width? 9 ½ feet at top (flare) 8 feet at the bottom
Crew? 4 men to handle boat & Cook.
Approximate years of operation, from? In 1700(s) to 1930.
Number of years you were active with the boats, including handling, building, etc.? 38 years.
Number of horses used each trip? 2.
Approximate load in high water? 8-10 tons.
Approximate load in low water? 4 tons.
Extent of trip, from Allagash to? foot of Chase Carry and return.
How often were trips made? Daily.
Time of beginning in the spring? Latter part of June. I would guess they ran until freeze-up in the fall.
Were there bunks on the boat? In cabin
Were the loads lashed down? No.
Did the horse or horses, ride back [to Allagash] on the boat? Yes.
Was the boat constantly snubbed coming down? No.
Was it ever allowed to run free? No, never ran free.
The boat was winched around Allagash Falls. By pulleys? Yes. What size blocks, single, 2 by 2, 3 by 2? 6 by double 2.
How many different hitches were made for this? Three.
Was the pulling done by men or horses? Horses.
What did the horses have on their feet, if anything special for shoes? Soft-iron shoes to stick to rocks.
Equipment on the boat and load? Pick-poles & rope & jill-poke (5 feet) to free from rocks we were caught on.
In barrels? Flour-200 lbs., & Beef-250 lbs., & Pork-300lbs, & Sugar-312 lbs., & Molasses-700 lbs. (called hog-set)
In bags? Oats-26 bg./ton & 13 bales hay/ton & beans 100 lb. bag & potatoes-100 lb.
In boxes? Shoes, rubber boots, cork [caulked] shoes—to go on logs, Canned food—beef, codfish, cabbage, turnips—Staple supplies are the main foods: also potatoes. Coffee and tea were drank. Not too much food was in circulation then—that was different times.
How long did it take to make one trip? To falls and back (July) one day & later in the fall two trips in three days. ([from] Falls to Long Lake - two day trip) Falls to Chase Carry it took four days to make a trip.
Did the boats take passengers, workmen, etc.? Usually men, but during hunting season hunters went and sometimes workers to woods and some took families.
What is the name for the short chain with ring on one end and iron on the other? Boom chain to hook boom (two logs) together.

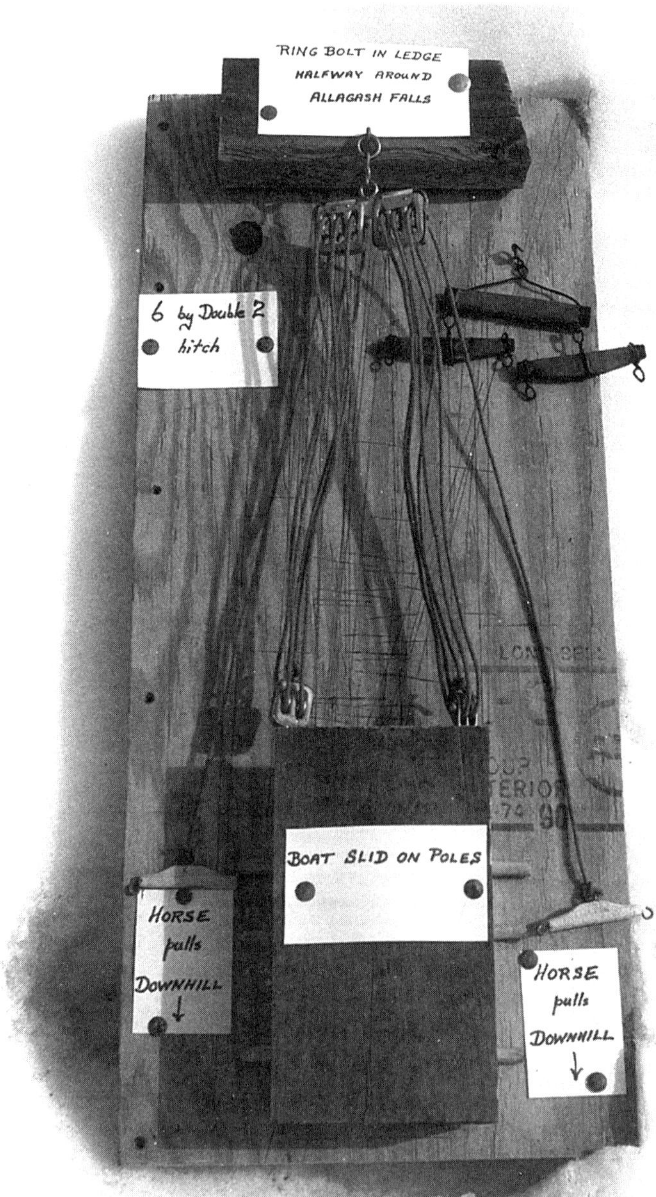

Norm Wight collection.

This model shows what that ring in the rock at the falls was for. The 6 by double 2 hitch provided enough mechanical advantage that only the two horses were needed to pull the boat up the hill.

[EPILOGUE]

WOOD HARVESTING TODAY

If you drive on some of the woods roads in the north woods, you may be surprised by the mountains of tree-length wood that line the sides of the road. The piles are 20 to 25 feet high and extend for two and a half miles or more. These piles are made up of more than 70,000 cords of wood! And this is a fraction of the winter's cut from just one woods camp.

The camps are made up of trailers—at first glance not much different from a regular trailer park, but these are not ordinary house trailers. The moderate-sized trailers serve as offices and bedrooms for foremen, clerks, and cooks. One may be a guest house. Large double-sized units serve as living quarters for the workers, and at the end of a snow-packed street you may find a building that houses a pool table and serves as a recreation room. Another double-sized unit serves as kitchen and dining room. A huge metal building serves as a garage where the woods machines are repaired. Somewhere nearby stands a small frame building in which a diesel engine hums continuously as it turns a generator which produces electricity for the camp.

Some areas are harvested mechanically, and some are done semi-mechanically. Mechanical means that the trees are cut and harvested by the huge woods machines. Where the terrain demands it, the trees are cut by men with chain saws and hauled to the yard by skidders. This is referred to as semi-mechanical. Many of the old-time woods terms survive. The man who operates a chain saw is known as a "chopper," just as Mushrat Murphy, who used an axe, was called a "chopper." The modern chopper's job is to cut the trees and fell them in such a way that the butts are heading in the direction the skidder will pull them. When the skidder arrives he hooks the trees to "chokers," small cables that slide on the large main cable of the skidder's winch. When several "stems," as the trees are called, are hooked, the skidder winches them to its rear, raising the butts off the ground, and then heads for the yard. Another kind of skidder is called a "grapple skidder." It has a huge claw on the back that grasps a bundle of stems and hauls them to the yard.

The chopper does no limbing except on trees that measure more than 16 inches in diameter. Limbing is done in the yard by a large machine called a Logma, which picks up the stem, whisks it at about 20 miles per hour through a circle of knives, and then places it on its pile, totally devoid of limbs.

In mechanical areas gigantic machines, whose tires alone dwarf a 6-foot man, are cutting trees. Two types are used. A machine called a Korin has gathering capability. It can gather six or eight cords of cut wood and haul it to the Logma. The other type simply lays the stems aside for the skidders to haul to the yard. Both harvesters use a hydraulically powered pair of shears to cut the stems, and they are capable of cutting up to an 18-inch diameter tree. The huge claw on the gathering harvester is capable of holding several stems at once before transferring them to the load. It works much like a person picking string beans, holding several in hand before placing them in the container.

Unlike the choppers, who work strictly piecework, machine operators are paid at an hourly rate. However, they do receive a bonus of so much per stem after 2,500 stems per day. According to a foreman, good operators make as much per day with their bonus as with the hourly wage. The competition for these high-paying jobs is fierce, and only a few qualify. There is only a 1-percent turnover of machine operators—a good indicator of the desirability of the position.

Women too have a presence in the north woods. "She's one of my best operators," the foreman says of their only woman machine operator. She not only operates her own machine, but is crossed-trained on the other machines too. Cross-training makes the operator much more valuable, because it is costly to have the expensive machines sit idle.

Not too many years ago, the paper companies cut for pulp alone. Today the high price of lumber makes it worthwhile for them to save their high-grade logs for the lumber mills. This practice has fostered more selective cutting. The more valuable trees, especially spruce, may be left to grow a little taller while the faster-growing and less valuable fir is cut for pulp wood.

PART THREE

Native Americans

Some stone artifacts found in the Allagash. The one at top left is a projectile point. The one on the right is probably a knife. The others are just waste from the manufacturing process. All are Munsungan chert.

PROLOGUE

There is good archeological evidence that there have been Native Americans in what is now called Maine for more than 10,000 years—since the end of the last ice age. The people of that time are referred to by archeologists as Paleo-Indians. University of Maine archeologists have identified a site near Munsungan Lake (just east of the Allagash) as a Paleo site. Here they find what were probably workshop sites, where the quarried materials were shaped into finished tools. The native people who have lived in Maine since the ice sheet receded have been given different names because their lifestyle changed over the centuries, just as ours have. Some believe they were several different people, but our present-day Native Americans disagree. So do I. They just changed with the times.

The people in this part of the book are of a time about 1,000 years ago—archeologists refer to this time as part of the Ceramic Period. The Native Americans in this part of the country were called Woodland Indians, and some researchers feel that these people were very late in their use of ceramics. The people belonged to many tribes, most of them defined and named by the river they lived near, but as a group they were called the Wabnaki.

Wabnaki, People of the Dawn, Children of Light; the northernmost people of the region we now know as New England. The Wabnaki tribes are known by the rivers associated with them, the rivers that give them life.

Everything in the Wabnaki world is sacred, because their own survival depends upon the survival of everything else. The animals are their relatives; the land is their mother. Plant and animal spirits are given thanks for giving strength to their bodies. Nature is their God, and their God gives them life.

The tribes of the Wabnaki move through their region in yearly cycles. In the early spring they move to the hills, where they will tap maple trees and boil down the sap to make sweet syrup and sugar. Later in the spring the tribes are at the falls of their rivers and streams, to catch fish with spears, traps, and weirs. The women gather groundnuts and greens, and dry and smoke the fish.

In the early summer these people move down river to their permanent villages, where they plant corn, beans, and squash. Then the families continue down to the river's mouth, where they hunt saltwater mammals, fish, and shellfish. From time to time, while they are at the coast, they will travel inland to care for the crops. These inland trips also give the opportunity to gather and dry berries.

Early fall signals the beginning of their move back upriver, where they gather medicines and wild nuts. The crops will be harvested and dried, some stored and buried in bark containers, and some carried with them. As the tribes continue to move inland they catch eels and hunt waterfowl, passenger pigeons, and other wild birds, as well as larger animals. This is a time to prepare for winter. The hides have to be tanned, clothing sewn, and meat dried.

By the time winter arrives the people will be in their traditional hunting grounds at the headwaters of their rivers, or sometimes beyond. Here they continue to hunt large animals like moose, caribou, and deer by overtaking them on snowshoes and killing them as they flounder in the deep snow. They also trap and hunt fur-bearing animals such as beaver, muskrat, and otter for food and clothing.

The members of the Bear Clan of the Penobscot tribe are moving into their traditional hunting territory now. This is their story.

[CHAPTER FIFTEEN]

THE BEAR CLAN

Running Bear stood in the canoe leaning on his long spruce pole as he waited for the rest of his band to catch up with him. He could not see them, but he could hear the swishing of their poles and paddles through the water as well as their grunts of exertion as they pushed their bark canoes against the current of Umbazooksus Stream.

Running Bear had long been the head of the Bear Clan, which he was now leading to the winter hunting grounds. His position of leadership was thrust upon him because of his natural ability to lead, not by ascension or popular acclaim. The others looked to him for direction and followed him. For his part, he accepted the responsibility without conscious thought that it should be any other way. His father, Standing Bear, had led the clan before him, but positions of leadership among the Penobscots were earned, not inherited, and while Running Bear had inherited the ability to lead from his father, he had not inherited the position.

The Bear Clan is part of the Penobscot tribe or, as they call themselves, The People of the White Rocks Country, or sometimes more simply, The People. The Penobscots are the largest of the Wabnaki tribes, and their winter hunting grounds extend not only into the headwaters of their own river, but beyond that into the headwater lakes of the Allagash. This is where the Bear Clan has its traditional hunting grounds.

Now, as Running Bear stood waiting, he thought about what lay ahead. Soon, today, they would reach the southeastern mark of the Bear totem—the beginning of his clan's hunting territory. He had spent many winters in these hunting grounds, some of them good, happy times when game was plentiful and there was plenty to eat. Memories of these happy, contented times were good. Other winters had been long, hungry times that seemed to drag on without end. These were times when he had to fight back tears as he helplessly listened to the children whimpering because their bellies hurt from lack of food. He remembered such winters during his own childhood. He remembered how he could not wholly understand why there was no food, all the while knowing it was the work of the great spirit, Manitou, and was not meant for The People to understand.

Other members of the Bear Clan were moving toward the hunting grounds by another route. The other band, led by Running Bear's cousin Walking Deer, would probably arrive at Sawanatkimoc* before Running Bear. They were traveling up Wassattegnetwick** to Kwanosa'gamaik*** where they would carry overland to Sawana'kimoc. The two bands would meet briefly at Sawana'kimoc before moving off to separate camps for the winter hunt.

There was good reason for moving and living in small groups. It was easier to find game to feed small groups spread over a large area than to try to feed one large group concentrated in one location. The clan carried jerked venison, dried fish, and parched corn, but these precious supplies were for hunters during the winter hunt or for survival food when game became scarce.

Now, as Running Bear stood in the warm, almost-still afternoon air, he wondered what the season would bring. *Would the animals allow themselves to be killed, or would they turn their backs on The People and leave them without enough food?* His thoughts were interrupted as the other two canoes came into view.

As he pushed off, sending his large canoe gliding easily upstream, a very faint odor caused an uneasiness within him. The smell was so quickly gone that he could not identify it, and he forgot it immediately.

With him were his wife, Sugar Maple and their two daughters, Wild Flower and Little Blueflower. These three, along with his son, Hawk-In-Flight, were the pride of his life. They were what made the difficult task of family and clan leadership worthwhile.

Not that Running Bear was alone in making his decisions. He had great respect for the wisdom and common sense Sugar Maple shared with him, especially during difficult times. Just as her physical strength was nearly that of a man's in paddling the canoe up the stream, so her inner strength was there to support him when times were difficult and the responsibility for others weighed heavy on his shoulders. As they moved the canoe up the narrow stream he admired her tall willowy figure, just as all who saw her admired it. Many of the other women of the tribe, her own age, looked at her with ill-hidden envy. She always held her head high with the assurance that comes with knowledge of one's own worth. Her wisdom was well known among The People, as well as by Running Bear.

Their daughters, now 11 and 13, were developing into independent women (sometimes to Running Bear's chagrin), who were certain to stand with their husbands as partners and not as mere drudges and bearers of children. This

* "South Lake," Telos Lake
** "Place Where They Spear Fish," East Branch, Penobscot
*** "Loon Lake," Webster Lake

trait they had surely inherited from their mother, and Running Bear knew it was good.

Coming around a bend in the stream, Running Bear suddenly saw the clear sky through the trees ahead, and he knew they had reached Umbazooksus Lake. The water splashing over the shallow riffles where the stream flowed from the lake told him it was time to go ashore and carry their possessions and canoes over the short trail to the shore of the shallow lake.

Just across the lake was the beginning of the trail to Megkwaka'gamik*. Here they would camp for a day or two. Running Bear eased the bark canoe up to the sandy shore. All four stepped out in turn, picking up as they did so an accustomed load to carry up to the shore. No one wasted movements or time on these carries. The procedure had been repeated hundreds of times as they traveled from place to place.

Each family member had his own Pitsonungan, or long pocket made of a small animal skin. In it they carried personal items and small tools necessary for their job in the family. In his Pitsonungan Running Bear carried stone and bone awls, knives, scrapers, and gouges he used to make snowshoe frames and canoes. He also carried his stone pipe, tobacco, fire-making tools, and a small bag containing the teeth of a bear, his personal medicine.

In their bags, Sugar Maple and the girls carried the knives, awls, and bone needles they used to sew skins for clothing, as well as a few personal items.

Running Bear sighed as he watched Sugar Maple shoulder a large bundle of items that couldn't possibly fit into her long pocket. These were her decorating materials, and it seemed to Running Bear there was no end to how many she collected. It wasn't that Running Bear didn't appreciate her talents. In fact, he wore much of her handiwork with pride. But it always seemed to him that she had more materials than she could ever use. However, he had long since learned to live among her collection and to hold his silence. Only occasionally, when he found himself rummaging endlessly through her materials looking for his own tools, did he complain.

In another large bundle were cured skins that Sugar Maple and the girls would use to make the family's winter clothing. There would be more skins later if the hunt went well. Several snowshoe frames were being carried to the hunting grounds. These frames would be laced later by Sugar Maple and the girls. Running Bear had made the frames during the summer when the family was downriver, where young, straight white ash trees grew. These trees provided the material he used to shape and bend the strong, lightweight frames. Some of the finished snowshoes would be used by the band during the winter, and some would be traded with other bands. Running Bear's skill as a canoe- and snowshoe-maker was well known among all the Wabnaki people.

* "Marsh Pond," Mud Pond

Other bundles in the canoe contained larger stone tools, cooking utensils, trap and snare materials, and various shaped bark containers. Of course, Running Bear's weapons were always within easy reach. His bow and arrows rode in the canoe directly in front of him. On the left side of his wide leather belt he wore his stone knife; on the right, his finely crafted stone ax.

Running Bear took pride in his carefully made weapons. He could afford to trade for the products of the best of the stone artisans, because he was the best of the snowshoe and canoe builders. One of the fine workers of stone was his ni-datbe, best friend, Grey Wolf. In a small bag in the canoe were blanks from Kineo, the stone mountain in the middle of Xsebem*. He had quarried these stone pieces for his friend as he and his family made their way up the great lake. He would deliver them to Grey Wolf when they visited his camp on Pongokwahemook**. Grey Wolf preferred these stones and those from Munsungun Lake for crafting fine weapons and tools.

Two large skin bags in the canoe contained dried corn they had harvested at the tribe's main village on the Penobscot. More corn was stored at the village. If the hunting went badly, some of the men would have to make the long trek through the winter snow to get the corn. It would let them survive, though barely, through the harsh northern winter. If the hunt went well, the dried corn they carried would be enough to supplement the meat, and would be carried as travel food for the hunters. The food stored at the village would be eaten in the early spring when they returned to the Penobscot.

Jerked venison and dried fish stored in bark containers would provide the clan with a varied diet through the coming winter.

Finally, there was one item in the canoe that Sugar Maple would trust no one to carry—her cooking pot. Running Bear had made a trade for this precious, fragile ceramic pot crafted by one of the tribes to the south. Cooking with it was so much easier for Sugar Maple than with the traditional birch bark containers that she could not imagine doing without it. Someday she would learn how to make these pots herself.

As the family returned to the canoe for their second load, the other two canoes came ashore to join them. First ashore was Running Bear's son Hawk-in-Flight and his uncle, Moose-Comes-Running. In the middle of their canoe was Hawk-In-Flight's wife, Autumn Leaf, and their son Little Fox. Running Bear quickly scooped up his 2-year-old grandson and set him ashore, where he momentarily admired the strong child as only a prideful grandfather can.

Running Bear's younger brother, Returning Thunder, and his wife, Sparkling Dew, paddled the last canoe. Their two small daughters stood in

* "The Extending Water," Moosehead Lake

** "Woodpecker Place," Eagle Lake

the middle of the canoe and smiled charmingly at their uncle as he helped them ashore.

In a short time the three canoes pushed off across shallow Umbazooksus Lake toward their camping place on the southeast shore—the beginning of the carry trail to Megkwaka'gamik. Again, for a fleeting instant, Running Bear's sensitive nostrils picked up the pungent odor, and again an inexplicable uneasiness came over him. He quickly looked at his uncle, Moose-Comes-Running, but if the older man had noticed anything his face didn't show it.

On a tree at the side of the trail where they landed was the totem of the Bear Clan. They were in their traditional hunting grounds again.

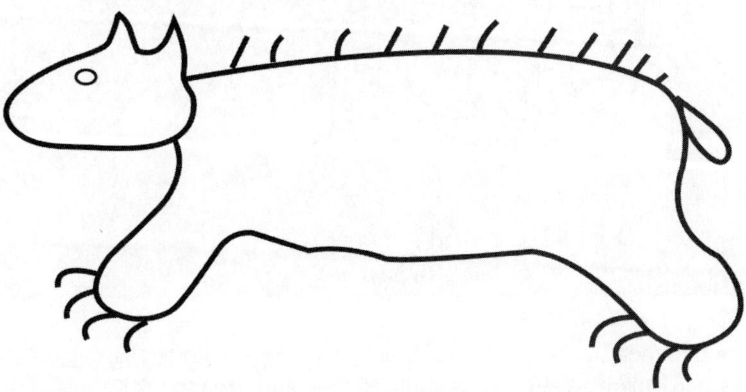

The men carried the canoes to the muddy shore of Megwaka'gamik, and Sugar Maple and the other women put the camp in order. While her daughters made necessary repairs to the bark lean-tos, Sugar Maple and her sister-in-law and daughter-in-law set out snares around the camp for rabbits and other small game. Then they took the children and gathered enough wood for cooking and warding off the chill that was starting to settle over them in the evenings.

Tonight they would eat fish that were caught back in Umbazooksus Stream, cooked with the roots the women had dug. Though each family would cook over their own fire, food would be shared equally. This was the Wabnaki way.

Sugar Maple arranged the stones around her fireplace and then hollowed out the soft earth in the middle of the circle of stones. Here she carefully placed her ceramic pot, which was bluntly pointed on the bottom. Gently she patted the soil around the pot so it would stand firm in an upright position.

After putting the vegetables and water into the pot, she carefully laid the fire around it. With a bow drill and its slotted base, Sugar Maple quickly produced a glowing coal, which she placed on some dried, shredded cedar and birch bark. By blowing gently on the glowing particle she soon had a small blaze, which she fed with progressively larger sticks of wood.

Sometimes The People carried their fire with them. Rotted hardwood trees, particularly yellow birch, produced the spunk. The spunk burned very slowly and almost without smoke. It burned so slowly that a small piece lasted half a day. Clam shells were lined with clay. A small hole allowed what smoke there was to escape. A small hollow held the smoldering spunk. The clay-lined shells were assembled, bound tightly, and carried in the skin of a small animal, such as that of a woodchuck.

The women cleaned the fish, skewered the heads to a slab of wood, and leaned the slab against a rock so the fish hung over the fire. The fish roasted with little tending while the vegetables boiled in the pot.

Sugar Maple assigned Little Blueflower to watch the fire and the supper while she and Wildflower spread the family's sleeping robes in their lean-to and arranged their possessions for safekeeping during the night.

It was almost dark when the men returned to camp. The family ate together, then sat around the fire. Gradually, the other members of the band joined them at their fire, and the children looked expectantly toward Moose-Comes-Running for a story before they went to their shelters to sleep.

Finally, Returning Thunder's daughter, Flower-Of-The-Pond, said, "A story please, uncle."

"Yes," said her sister, Sweet-Grass-Swaying, "tell us how the great giant-hero Gluskabe came to the Wabnaki people."

"And why is he called 'The Liar'?" Flower-of-the-Pond asked.

The old hunter sat quietly staring into the fire for a few moments. Finally, he began.

"Originally there were two giants who lived with their grandmother Poksinsques. They were twins, and they grew up together until they were gigantic in size. They looked so much alike that even the grandmother could not tell them apart. But though they looked exactly alike, no two people could be more opposite. One brother was a genius for creating all manner of things that were good, while the other brother was equally inventive in creating evil. One brother was smart and clever, and the other was stupid. One brother was open, outgoing, and kind. The other was selfish and cruel. Any characteristic that could be found in one, the opposite was in the other.

"One day the good brother decided to make some fish. He molded them out of dirt, breathed on them to give them life, and placed them in the water. At first the fish he made were crude and not too attractive, so we have some ugly fish, but he continued to make the fish and eventually they became more and more beautiful. Finally, he fashioned the most beautiful fish of all and breathed on it to give it life. But before he could place it in the water the other giant came along, snatched the fish from him, and stepped on it, making it flat. Now instead of his very best fish, all that was left was the flounder. The good brother was very discouraged by this, and he gave up making fish. The world, it seemed, was not big enough for both brothers.

"A little later the bad twin asked his brother if he could be killed. If he knew his vulnerable spot he could, he explained, protect him from harm. The good brother immediately replied, 'To kill me, one must hit me on the head with a cat-o-nine-tails. Now you must tell me how you may be killed so I may protect you.' To this the stupid evil twin replied: 'To kill me, one must hit me on the head with a stone axe.'

"That night the good twin hid his great stone axe beneath his sleeping robe when he went to sleep. After all were asleep in the wigwam, the bad brother crept from his robe, seized the cat-o-nine-tails, went to where his brother slept, and struck him over the head as hard as he could. The good brother immediately hit his evil brother over the head with a great stone axe and killed him. So it was that Gluskabe (The Liar) got his name. With the death of the bad twin, the genius for creating evil was gone. But today The People must deal with the evil that had already been created."

A faint trace of a smile played on the old hunter's face as he finished his story and said, "To sleep now, tomorrow we must hunt."

As the children went with their mothers to their sleeping robes, Running Bear, Moose-Comes-Running, Hawk-In-Flight, and Returning Thunder talked about what they would do in the morning. It was true, they did need to hunt. Food was running low, and it wouldn't do for them to start eating the precious dried food they carried for the long winter ahead.

So they decided that in the morning the hunters would carry to Megkwaka'gamik items not needed in camp. Hawk-In-Flight would return directly to the camp to stay with the women and children. The other men would hunt back toward camp, each making a huge arc on either side of the carry trail. If the hunt was successful and if the women's snares brought them game, they could continue moving toward Sawana'kimoc the next day.

Sawana'kimoc was the southernmost lake of the Bear Clan's hunting grounds. The Wabnaki hunted one third of their land each year. Of the one third that was hunted, they only took two-thirds of the animals, leaving the remainder to repopulate and grow. This careful management of these gifts assured The People of continued well-being.

The men knew they were not late getting to their hunting area and there was really no hurry. Yet each considered the hunt, and the hunting grounds, very special and sacred. They could not help being eager to get back to the land they loved best. The hunt was special—a time for the utmost of patience and skill. Even when Manitou was smiling and there were plenty of animals, they were not just handed over to The People. The hunter had to hunt. That was the way it was suppose to be.

The men rose as one to go to their sleeping robe. But before turning in, Running Bear walked around the outskirts of camp. He stopped frequently, looking, listening, then continued to walk. All was secure. No large animals were near and the Mohawks, "The Cannibals," traditional enemies of the Wabnaki, had not been in these parts for a long, long time.

Suddenly Running Bear heard a thrashing in the brush nearby. His hand instinctively went to his stone ax as he stared through the darkness. Finally, he saw the cause of the noise. One of the women's snares had caught a rabbit. *Perhaps a good omen for tomorrow's hunt,* Running Bear thought. He quickly killed the animal, reset the snare, and returned to the camp, thanking the rabbit-spirit for his body as he walked.

As he slipped into their sleeping robe, Sugar Maple drowsily placed her hand on his shoulder. He touched her lightly in return, and soon they were both asleep.

[CHAPTER SIXTEEN]

HAWK-IN-FLIGHT'S MOOSE

It was early in the morning when the four men shouldered their loads and headed toward Megkwaka'gamik as planned. Characteristically, these carries were made into a race, with each participant boasting of his ability to carry a heavier load faster than any of his companions. Good-natured jibes were hurled back and forth. The first to reach the end of a carry rearmed himself with a fresh supply of verbal ammunition, which he brandished unmercifully upon his fellow carriers. When a group of young men were traveling together the competition was especially fierce, but even with carriers of unequal age the game was played to distract the mind from the tedious chore ahead and the aching muscles that would result.

Nor were practical jokes unheard during these sometimes long, difficult carries. A joke good for hours of laughter as it was told and retold was to slip heavy stones into the load of an unsuspecting carrier. This trick, which couldn't be used too often, was reserved for a young man who won more than his share of these races and did more than his share of boasting of superior strength and speed. The trick was successful if the rocks were carried to the other side. It didn't matter if he won that day or not. Of course, it was all the better if he lost the race, but even if he won, the fact that he was foolish enough to carry the worthless rocks made him the brunt of the others' jokes for a long time—until a better trick was played on someone else.

Today, as was usually the case, Hawk-In-Flight was the first to reach the canoes hidden on the shore of Megkwaka'gamik. He always teased his elders about their slowness in a respectful way. Today it was even more subdued than usual, because he had to return to camp instead of hunting with the older men. He knew someone had to stay with the women and children during the day, but he could not help envying his father and uncles for the day of hunting they would have. He didn't feel as if he were being treated unfairly, but neither could he find much to be cheerful about.

The four men stored their loads where they would be safe overnight, then sat by the pond, talking as they rested and loosened aching muscles. Shortly, Running Bear, Moose-Comes-Running, and Returning Thunder moved off to the sides of the carry trail, starting their daylong hunt, which would take them back to camp by evening.

Hawk-In-Flight continued to sit by the dark waters of Megkwaka'gamik, delaying for a few more moments the inevitable return to camp and a long boring day ahead. As Hawk-In-Flight watched the mist rise from the surface of the pond, he became aware of a faint splashing sound to his right. The shoreline was hidden by trees, but as he listened the splashing grew louder and he knew it was a large animal, probably a moose. His hunter's heart quickened as he scooped up his weapons and silently moved to conceal himself. His heart continued to pound in his chest as he crouched in the underbrush, straining his eyes to see through the leaves that were just now starting to turn their autumn colors. For what seemed interminable periods of time the splashing would quiet as the animal stopped to feed or survey the area for signs of danger. Hawk-In-Flight knew that if the animal continued to move along the shoreline it would pass within a few paces of where he was hidden. He checked his bow and arrow held at ready. He carefully moved each of the arrows in the quiver to insure they could be pulled free for a quick second shot.

For Hawk-In-Flight the thrill of the hunt was made all the sweeter by the fact that he, who was chosen to return to camp, might be the one to bring in meat after all. He thought of these things as he waited, but gradually his excitement began to abate as he realized it had been a long time since he had heard a sound. He had to stop himself from running to the shoreline to see if the animal was still there. But the years of training by his father and uncles in the value of patience prevailed, and he continued to wait.

Then a movement through the trees! Black! A splash! It was a moose, and it was moving to within a few paces of Hawk-In-Flight's hiding place. The long period of silence must have been caused by the moose swimming silently across a deep spot in the pond. Hawk-In-Flight slowly raised his bow to the ready. If a stray breeze did not waft his scent to the moose, he would have his chance momentarily. The moose moved into full view and paused. A young cow—barren—no calf. Hawk-In-Flight drew his bow, aimed at the huge chest, and released the arrow. The moose jerked its head upwards and hunched silently as the stone-tipped arrow buried itself in its chest. Hawk-In-Flight wondered later if the animal reacted to the sound of the arrow whispering through the air or to the actual impact. Another arrow quickly fitted and released struck a hand's width from the first. This time the animal bolted, but frothy blood was pouring from its wounds and from its nose. Hawk-In-Flight knew the animal would not go far. He watched as it ran down the shore to his

left, then heard it breaking brush as the cow turned away from the pond. Then all was silent. The frothy blood meant the moose was shot in the lungs and would die quickly.

Hawk-In-Flight stretched his cramped muscles, then relaxed. He would wait. The wounded animal would not travel far before it stopped, if there was nothing to frighten it. With luck it would not be far away.

When he felt enough time had passed, the young man moved down the shore and had no trouble picking up the blood trail. He saw that the large animal had stopped running almost immediately after entering the woods and the blood signs became heavier and heavier. As he moved silently along the trail he glanced ahead looking for the black hulk that would be the dead moose. In a short time he saw the animal stretched out on the forest floor.

He approached cautiously. More than one hunter had been badly hurt or killed by an animal presumed dead. With the tip of his arrow he touched the eye of the moose. No reaction. The animal was dead.

Now the thrill of the hunt was replaced with an unexplainable sense of remorse—a feeling shared by all true hunters, one that reminds them that they, as well as their prey, are part of the mother earth. Hawk-In-Flight knelt and thanked Manitou for providing his family this great animal, and he thanked the spirit of the moose for allowing itself to be killed. He told the moose-spirit how its body would become part of the bodies of members of the Bear Clan, mightiest hunters and fighters of all the Penobscots.

His prayers over, Hawk-In-Flight set about the business of caring for his kill. After removing the entrails and opening the chest, he propped open the body cavity with sticks to let it cool. Then he sat back to decide what to do next. Certainly he could not handle such a large animal alone.

He decided to return to camp since he was already late. He would get the women and children to help him, or perhaps wait for Running Bear and his uncles to return.

His spirits soared during the walk back over the carry trail. He was providing enough meat to last many, many days. He was eager to tell Sugar Maple of his adventure, but even more important would be telling his tale to his father, uncle, and great-uncle. For while he knew in his heart that he could not yet equal the hunting skill of these great hunters, he was thankful for the opportunity to be worthy of their praise. With the cool weather now coming to the north country, this meat would last them through their travels to the family's permanent hunting camp. They would no longer have to delay their travels to hunt for food.

When he had returned to camp on Umbazooksus, Hawk-In-Flight told of his good fortune, but he omitted the particulars. These he would tell later, when the whole band gathered around the campfire. Sugar Maple showed her pleasure by promising him a meal of baked moose nose, a great delicacy. Hawk-In-Flight knew that traditionally the largest portion would go to him as the successful hunter. Right now though, he had to talk to his mother about what to do next.

For Sugar Maple the logical solution was obvious. With an ample supply of food, there was no reason to tarry at Umbazooksus camp. They would pack the remaining gear and move to a similar camp on the shore of Megkwaka'gamik, where they would begin immediately to care for the moose carcass and pack the meat for travel.

In a very short time the band, accustomed to such moves, had everything packed and ready to go. Before leaving, Sugar Maple left a message on a piece of birch bark for Running Bear and the others.

By mid-afternoon Hawk-In-Flight, the women, and children had set up camp at Megkwaka'gamik, and everyone was hard at work caring for the huge carcass. Nothing would go to waste. Sugar Maple, Autumn Leaf, and Sparkling Dew had removed the hide and laid it out flat to provide a clean surface upon which to cut the meat. The women's only tool for this job was the ulu knife. This half-moon-shaped stone tool was efficient and simple. With it they could skin an animal, dress the hide, cut meat, disjoint a carcass, and perform many other domestic chores that required a cutting tool.

While the women worked on the moose, Hawk-In-Flight gathered birch bark to make containers for the large supply of meat. As he moved through the forest collecting the bark and rolling it, he thought about how important this bark material was to all of the Wabnaki people. The story of Gluskabe and the birch tree came to mind.

Gluskabe named the birch tree and asked it to take care of all the Wabnaki people. One day he found a straight birch tree and wanted its bark to make a canoe. When Gluskabe cut the tree, it fell in such a way that he was almost killed. This narrow escape normally would have provoked enough anger, but Gluskabe suffered the additional indignity of having an exceedingly difficult time extricating himself from the tree's branches. Enraged, Gluskabe took up a stick and beat the birch from one end to the other as hard as he could. Then he commanded the marks he had made on the white trunk to remain as eyes, so that the birch could see where it fell and so never kill or almost kill again. The People still call these marks birch tree eyes. Some are long eyes and others are short eyes.

Shortly after sunset Running Bear, Moose-Comes-Running, and Returning Thunder approached the newly made camp. They had seen Sugar Maple's message, so the appetizing smells of liver, heart, and moose nose cooking came as no surprise to them. They were empty-handed, but smiling broadly at Hawk-In-Flight's good fortune. As final preparations for the evening meal were made, they told of their rather uneventful day. They looked forward to hearing about Hawk-In-Flight's kill.

After the delicious supper of freshly killed meat, the entire band gathered around the fire and Hawk-In-Flight told his story. No detail was omitted, and no one spoke until he was finished.

The children had seen moose before, of course, but this was the first one they had actually helped to care for from start to finish. Having observed it close-by this time, they were surprised at its unusual features —not at all graceful in appearance like most other animals. Moose-Comes-Running shifted his position slightly, and they all knew a story was coming. The children looked at him eagerly, their eyes shining in the firelight.

"You remember I told you how Gluskabe originally made some of the animals far too large for man to manage?" he said.

The children nodded.

"Well, Gluskabe examined the moose, and he knew it would prove far too much for man to cope with, so he called to the moose, held out his hand, and said, 'Push against this!' The moose did, and pushed so hard that he shortened his neck and compressed his shoulders. He continued to push so hard that he broke his nose, but by now he was reduced to manageable size, so Gluskabe allowed him to stop. To this day the moose has a broken nose, a short neck, and humped shoulders. And, if you look very carefully, you will see in the fur on its head the imprint of the creator's hand."

The children, as one, jumped to their feet and raced to the head of the moose, which had not yet been cared for.

"There it is!" said Little Fox, "There it is, I see it!"

As the children returned to the fire, Wild Flower, who had heard the moose stories before, egged the old story teller on by asking, "But if Gluskabe made all of the moose small, then where did we get so many small ponds that are said to be the hoofprints of a giant moose?"

As expected, with no coaxing needed, Moose-Comes-Running went on with the story.

"It seems that one moose escaped the pushing contest and remained its original size. This huge animal became a great burden to The People as he rushed through villages knocking down wigwams and creating havoc in general. When Gluskabe heard of this, he went out and searched for the creature. He armed himself with his giant axe and took along his little dog. Gluskabe and the dog had an agreement: man would feed the dog, in return the dog would serve man. After a long search Gluskabe found the giant moose and started him running. Each step of the moose made deep imprints in the earth that filled with water, so now there are whole chains of small ponds connected only by a small stream."

The tale was interrupted momentarily as the older man accepted the stone pipe Running Bear had lit and offered to him. After a few puffs he continued.

"The great moose ran for four days and nights, but on the fourth day it started to tire. Gluskabe continued his pursuit as fresh as when he started, leaping valleys, vaulting mountains, now gaining on his quarry. On the fifth day Gluskabe caught the giant moose and swung his giant axe with deadly accuracy.

"Gluskabe was hungry, so he dressed the huge beast and threw the guts to his dog. Part of them fell on land; the rest went into the Great Salt Water. The dog ate what fell on land, but would not touch that which fell into the water, so Gluskabe turned the dog to stone on the shore. The guts turned into white stone in the water. The stone dog can still be seen looking into the water at the white stone entrails.

"The rest of the moose Gluskabe cooked in a pot and ate. Then he walked inland and overturned the pot, and it became Kineo at the great lake, Xsebem."

[CHAPTER SEVENTEEN]

THE BURN

The chore of properly caring for the large moose carcass was not an easy one. Bark containers had to be fashioned and the meat carefully packed into them. The hide had to be thoroughly scraped to remove all traces of fat and meat. Then, as long as the weather remained cool, it could be rolled and carried to the permanent camp before being stretched to dry. The brains would be saved for use later in tanning.

Running Bear announced he was going to take a canoe and go out to A'pmoojenegamook*. He wanted to look for some good cedar for canoes, he said.

Sugar Maple smiled and nodded. She knew her husband well. He was always looking for canoe materials, but the real reason for going was to get away from all of the domestic activity that would be going on around camp today.

Little Blueflower and Wildflower, almost in unison, asked their father if they could go with him.

"No, I think your mother needs you here," he said.

They looked at Sugar Maple, and she paused at the fire a moment. Of course, it was important for the girls to learn the work of the women so they would become good wives and mothers. However, it was equally important for them to learn the ways of the wilderness and how to survive in it. Pleasant memories of times spent with her father flashed through her mind. What he had taught her had helped to make her the self-sufficient person she was today. She knew she could spend days in the woods alone if necessary. Yes, these times with their father were important for her daughters.

"They both worked very hard yesterday," she said. "It will be all right if they go."

"But," she added quickly as the two girls scrambled to get their things, "I expect you to bring back some plant-food to go with all this meat we have."

* "Cross Lake," Chamberlain Lake

It was only a short paddle across Megkwaka'gamik to the small stream that flowed into A'pmoojenegamook. When they arrived at the small outlet, Running Bear carefully beached the large canoe and stepped ashore to look the stream over. There was not much water flowing from the pond, and the stream was low. They would line the canoe through to A'pmoojenegamook, walking along and guiding it between the rocks.

This small stream, like the carry between the ponds, had been used by countless generations of Running Bear's people. Through the many seasons of use they had picked out rocks that were in the way until, now, the stream almost had the appearance of a man-made canal.

As they walked along, guiding the canoe carefully between them, Running Bear talked about the area and pointed out trail signs that marked the families' land routes through the huge hunting area. Stopping to move a large rock that had probably been rolled into the stream by winter ice, Running Bear spotted something dark moving in the dense evergreens nearby.

The movement was caused by a small flock of spruce grouse. These stupid birds were almost always easily approached, and Running Bear saw the chance to show his daughters an important survival trick. While the girls watched, he fashioned a noose from a piece of line he had taken from his long pocket. The line, made from the bark of the basswood tree, was intended for fishing, but it would serve this purpose well too.

Next he attached the noose to a dead alder branch in such a way that when the line was pulled, the noose would close while still remaining on the pole. He told Little Blueflower to hold up her finger, and when she did he deftly dropped the noose over it and pulled tight.

"Now," he said, "take it and catch a spruce grouse."

Little Blueflower smiled in anticipation as she took the snare and moved cautiously toward the feeding birds. They ignored her approach, and she was soon standing close enough to try the noose. On the first try the line caught on a small branch, which she failed to notice because she was so intent on the grouse. The branch swung out and hit the grouse in the breast. The bird fluttered to another branch and sat there looking at the three people. Little Blueflower tried again on another bird, and this time the noose dropped over its head and she quickly pulled it tight.

The grouse fluttered madly and Running Bear quickly grabbed it, loosened the noose, and then let the bird go. "We don't need meat now," he said.

In spite of all the frantic activity the flock of grouse was still in sight, so Wildflower took the noose, and soon she too had her quarry fluttering on the end of the pole. Again, the bird was released. Running Bear said, "I wish their sweeter-tasting cousin, the ruffed grouse, was that easy to catch."

As they walked back to the canoe Running Bear rolled up the basswood line and put it back into his long pocket. "Do you know why we call the ruffed grouse Mitchiwess?" he asked.

"No," Wildflower replied.

Running Bear squatted down by the canoe and told his daughters a story about the ruffed grouse. "Mitchiwess means canoe-builder-for-all-the-animals. The ruffed grouse got this name because his drumming in the woods sounds like a canoe builder working, as you have heard me doing many times. Mitchiwess decided that he would build a wonderful canoe. He reasoned that if a canoe had two ends and could go two ways, he could build a round canoe that could go all ways.

"Mitchiwess called forth all of the tribes—the bear, the woodchuck, the owl, the bluejay, and the porcupine—to watch the first run of his wonderful canoe. When Mitchiwess got into the canoe and paddled it, all he did was go round and round. He was so ashamed and embarrassed that he went back to the deep woods to stay. That is why the partridge is so shy and wary today."

The three continued walking the canoe along the small stream, stopping occasionally to roll out a boulder that blocked the way. Running Bear remarked it was likely that no one had passed through the stream this season, as the rocks from the previous winter were still in the waterway.

Soon they were stepping out of the woods and gazing across the expanse of A'pmoojenegamook. Now Running Bear felt as if he were back in his beloved north country again. It wasn't until he saw the sunshine on the lake and felt the breeze of open water that he felt this way. Released! Free again, free from the closed-in feeling of the dense forest that lay just behind him. As they got back into the canoe, he noticed the breeze was starting to shift toward the south. If it continued to move around they might be in for rain.

They paddled out into the huge cove, and though no one spoke of it, there was a mutual desire to get out of the cove to where they could see the lake stretching out to the north and to the south. Far to the south they would see the great mountain Ktaadin rising majestically above the forest.

As they paddled, generally following the right shore, the wind continued to swing to the south, and Running Bear suddenly caught a whiff of that pungent odor whose identity had eluded him twice before. This time its source was very clear. Forest fire! But it was not a fire in progress that caused the odor, but the smell of burnt land. Running Bear knew from the direction of the wind that the fire must have burned part of his clan's hunting area, the section they were to use this year! This thought troubled him, but he had to resist the urge to go immediately to survey the damage. The trip was too long for today. They were expected back at camp before nightfall.

Out on the great lake Wildflower and Little Blueflower smelled the burnt land too. Wildflower asked, "What will we do if our hunting camp is burned?"

"We can rebuild the camp," Running Bear answered, "but if the area destroyed was large there will be no animals to hunt. We will have to go to another area. We will see." Running Bear went on to explain that if the fire had been small, it would actually be beneficial to The People and to the animals. The fire would kill all of the big trees, and the small plants, food for the animals, would receive the sun they needed to grow. If the area of the burn was large, all of the animals that escaped the fire would not be back for many, many seasons—until there were trees big enough to hide them from their enemies and shelter them from the cold north wind of winter.

As they paddled the canoe back toward the shore, Running Bear thought about the rain that he was sure was coming. He had a plan for using the rain to help his band move its heavily laden canoes into A'pmoojenegamook. At its present level the small stream connecting the two bodies of water was not deep enough to float the loaded canoes. On the north shore of the cove was a small stream that drained a small pond* into the lake. Running Bear and his daughters landed their canoe near this stream.

* Lost Pond

"Here is a good place for you to gather the vegetables your mother wants," Running Bear said. " I will look for some cedar." He pointed in the direction of the small pond and said it was only a short walk away. He instructed them to confine their search to the area along the stream's course and to meet at the canoe by the time the sun was halfway to the western tree tops. Then he turned and left them alone.

The girls found wild chives growing on the rocky beach where they had landed. They gathered a supply of them, relishing the thought of the stew their mother would make using them as flavoring.

Leaving the chives at the canoe, they moved inland. Soon they spotted a familiar-looking flower. Though dried and withered, they knew it to be the sunflower potato. They had helped to cultivate it back in their village on the Penobscot. The tubers of this plant were delicious this time of year, and they gathered a large supply. On their way back to the canoe they decided to gather some bunchberries while waiting for their father to return. The bright red berries were not really plentiful here, but the sisters managed to gather several handfuls, not counting those they ate.

When Running Bear returned he told them how these bright berries could be used as bait to catch small fish, which in turn could be used to catch larger fish.

Shortly before sunset, the three were back to where Megkwaka'gamik Stream left the pond. Running Bear told his daughters about his plan. "Many times we have broken open the dam of the beaver to provide enough water to use our canoes on small streams. With the rain coming there will be much water. We will build a dam, like the beaver, to save some of it. Then, when we are ready, we will break open the dam and there will be water to float our loaded canoes."

The project appealed to the two girls. It was something new, this dam building like the beaver, and they could perhaps avoid the toilsome job of carrying everything between the two bodies of water. First they collected rocks and large sticks, which they piled across the outlet. Next they carried armfuls of forest litter, leaves, small sticks, and soil to the dam and stuffed them into the openings. Finally, they placed a layer of litter on the pond side of the dam.

"We would not make very good beavers," Running Bear said, "but I think it will stop some of the water."

The rain came that night and lasted through the next day and night, finally clearing and turning colder.

The dam worked well, raising the water level of Megkwaka'gamik to the point where the water started washing around the dam. The water that had been stored was released into the narrow stream and floated the loaded canoes even better than Running Bear had hoped. They could not get into the canoes, because the narrow stream allowed no room to maneuver, but little effort was needed to walk alongside and guide them in the swift-moving water. When they reached A'pmoojenegamook they paddled out of the cove and made camp on the point of the land where there was a good view up the lake.

The men had discussed the possibility of the hunting grounds being burned, so they agreed that Running Bear and Moose-Comes-Running would paddle down to Sawana'kimoc to survey the area and try to make contact with Running Bear's cousin, Walking Deer. When they met they would have council and decide what the clan was to do.

As was often the case in this country after a rain storm, a hard northwest wind was blowing on the morning that Running Bear and Moose-Comes-Running left for Sawanatkimoc. They had some difficulty launching the canoe through the heavy surf on the gravel beach, but once out on the lake they expertly swung the canoe southward and were soon out of sight of their camp.

It seemed that almost no time had passed before they were opposite the great arm of the lake that reached out to the east, and soon thereafter they entered the thoroughfare that connected A'pmoojenegamook with Pataquongamis*. As they entered they saw the burn for the first time.

The fire had destroyed everything up to the thoroughfare and to the south as far as they could see. It had also burned an area to the east toward Kwanosa'gamaik, where the rest of the clan planned to enter the hunting grounds.

Moose-Comes-Running said, "From this time on this place shall be known as Pataweekongo'moc, 'Burnt-Land-Lake.'"

Running Bear agreed it was appropriate and said, "We must learn where Walking Deer is camped. They must have reached this lake by now. Let us go to where the winter camp once stood to see if there is a message there for us."

The two men paddled to the south shore of the lake, and as they approached the charred remains of the clan's camp they could see a piece of bark fluttering on a blackened stump. A charcoal message on it was from Walking Deer. It said the band had moved through the thoroughfare and was camped on the southeast shore of A'pmoojenegamook, where it would stay until Running Bear joined it.

Running Bear was surprised they had not seen the camp as they came down A'pmoojenegamook, but Moose-Comes-Running reminded him that they had been concentrating on handling the canoe in the large waves. It was likely that Walking Deer's band had seen them, however, so they would be expected.

The wind had gone down somewhat by now, and the two had little difficulty paddling back on Pataweekongotmoc and through the thoroughfare.

*Round Pond

The waves were still large on the great A'pmoojenegamook, but the two men expertly quartered the large canoe into the waves and were soon within sight of Walking Deer's camp. Walking Deer had been watching for them, and he and the rest of the band greeted their relatives warmly. It had been many days since the two groups parted back on the Penobscot, and there would be much to talk about.

The women prepared a huge meal, and everyone ate well and enjoyed the reunion. After the meal they gathered around the central fire, as was the custom, and Walking Deer couldn't resist the opportunity to tease Running Bear for passing his camp without seeing him, though he knew the reason for it.

Running Bear took the goading good naturedly and smiled somewhat sheepishly.

Moose-Comes-Running said, "It was the wind-bird acting up."

All of the children knew Moose-Comes-Running to be a great storyteller, so several quickly asked about the wind-bird.

The adults smiled too, for they also enjoyed the old hunter's stories, especially the ones about their folk-hero, Gluskabe.

Moose-Comes-Running settled himself comfortably and told the story. "One day Gluskabe was out in his canoe in the wind, much like we were today, except he was paddling against it instead of with it. In those days the wind blew continuously. Greatly irritated by the wind, Gluskabe traveled to where the wind-bird sat—having first learned of his location from his grandmother. On the way the wind blew so hard it blew his hair off, which did not help his disposition any. When he reached the bird, Gluskabe picked him up on the pretense of moving him to a better spot, but instead threw him down, breaking a wing. Satisfied with his handiwork, Gluskabe went home, but now the wind never blew at all—it was unnaturally still. Even worse, the sea became covered with scum and began to stink.

"And, there was no relief from the black flies and mosquitos. So Gluskabe returned to the wind-bird, who did not recognize him because his hair had grown back. Gluskabe realized that the bird had been doing what was best for man after all. The great giant was sorry he had broken the bird's wing, so he healed it and put the bird back on its perch. But he also persuaded him to moderate his wind-making. The bird agreed, and since then the land has had gentler winds interspersed with absolute calm. So, you see, if not for Gluskabe, Running Bear and I could not have been on the lake at all today, so great would have been the wind."

The younger children's heads were nodding sleepily by the time Moose-Comes-Running finished and their mothers moved to put them to bed, but the older ones asked for one more story. The old man said he would tell a short one and then they must all go to their robes.

"This story will give you something to think about while you wait for sleep to come," he said. "The story is about how Gluskabe altered the seasons. In those days the world was covered with ice and snow the year round. The giant is said to have tricked the keepers of summer and then run away with the summer jelly, placing it before the house of the chief of Winter, whom he threatened to melt if he did not moderate the cold. In this way the four seasons came about, and The People had ni'ban, summer; dagwo'gawi, fall; pabotni, winter; and si'gwan, spring. To bed," he said as he folded his arms across his chest, indicating he would say no more.

With that final note, the women and children rose and went to their shelters for the night. Walking Deer put more wood on the fire, and the men settled down for council. The women who wished to do so would return to take part in the meeting. They had to decide soon where the clan would winter. Their traditional hunting area extended from the land south of Pataweekongomoc up the western shores of all the Allagash lakes, to where the Allagash River itself tumbled over the rocks as it left Allagaskwigam'ook*. The total area was large, extending westward from the lakes about one day's walk. Boundaries were naturally occurring features like streams and ridges. At each crossing was posted the clan's totem.

* "Bark Cabin Lake," Churchill Lake

[CHAPTER EIGHTEEN]

REUNION WITH GREY WOLF

The problem of settling on the clan's hunting area for the winter was of paramount importance right now. Not only was it important for them to ensure a good supply of game for their families' food and clothing, but also they were responsible to their mother, the earth, for not killing too many animals in any one area. The men in the council discussed this.

Running Bear said, "The area we hunted two winters ago is large and should have many beaver. There was still much game when we were there last."

Moose-Comes-Running agreed and added, "It is possible we can extend our area to the lake called Chemquasabamticook. Our totems are now one half day east of there, and no one uses the land between."

"Our brothers the Walastukwiak*, the Malacite, hunt the area from Wallastook** to Chemquasabamticook," said Running Bear. "Are their totems on the shore of that lake?"

Walking Deer said, "They are. I traveled through that area several seasons ago to Wallastook. I remember there were many streams with beaver lodges in them."

Among all of the Wabnaki people, land was held by a family or clan as long as it was used. If an area went unused, then it could be used by anyone who had need of it. Such was the area that the Bear Clan was considering.

The presence of beaver in the hunting grounds was of paramount importance. Without this hard-working animal, the trees formed an almost unbroken canopy over the forest floor. The dense shade from this canopy did not allow the low-growing plants to live, and so there would be few, if any, animals. But where there were beaver there was never-ending change. Old trees died as the dammed-up waters spread. New vegetation appeared around the edge of the marsh and attracted animals of all kinds. First the plant-eaters, then the meat eaters who would keep the others in balance. Then, when the beaver were finished with the area, for whatever reason, the water would

* "People of the Beautiful River," Malacite

** "Beautiful River," St. John River

drain from the marsh, leaving a fertile meadow that supported a large variety of plants and animals, until the meadow, too, was finally taken over by the mature trees and again the dense canopy of leaves closed in and choked off the life below.

Not only did the beaver's work benefit all of the animals upon which The People depended, but the beaver itself was their staff of life. The flesh was used for food, the fur for clothing, bed coverings, hangings in the wigwam, and gifts. Beaver robes were highly esteemed as a marriage dowry. A substance called castoreum, a secretion from the scent gland of the beaver, was used to remedy earache, deafness, headache, loss of memory, and insanity. The skins were a medium of exchange and highly valued by all of the tribes. So The People used the beaver wisely. They well knew that if allowed to reproduce unchecked, the beaver would eat all of its food and soon have to move to a new area. The trees would take over again, and in a short time there would be no animals to feed and clothe The People. The Bear Clan's area was large enough that there was no danger of all the beaver moving out, but still, the beaver was a gift from their Manitou and had to be used properly.

"It may be possible for us to use our usual hunting camps on Pongokwahemook or Allagaskwigamook," Walking Bear said, "but, if not we should start building new wigwams in a new location as soon as possible. " Each man told what he knew of the area, and it was late when they finally decided on a plan to scout the area—a plan that would allow them to be quickly done with the scouting and still see to the safety of their families.

Tomorrow Walking Deer and his band would pack up and move up A'pmoojenegamook, through the small stream* connecting that lake to Pongokwahemook, and then on to Grey Wolf's camp near Nahmajimskit'egwek**. Running Bear's band would join them as soon as possible. The women and children of both bands would remain with Grey Wolf while the men scouted the new area. Running Bear looked forward to seeing his old and best friend again, but wished he could spend more time with him. Anyway, he could deliver the stone blanks from Kineo, and they would get together again later in the season.

Running Bear, Moose-Comes-Running, Hawk-In-Flight, and Returning Thunder would move up the Allagash stream to Allagash Lake. They would scout north from there to the southeast shore of Chemquasabamticook. If the weather remained mild, much of their traveling could be done by canoe. If the weather turned cold, however, the small ponds and marshes would freeze over quickly, and it would be necessary to move about on foot.

Walking Deer, his two sons, and his younger brother would travel up Pongokwahemook, through the thoroughfare to its northern end where

* Present-day Lock Dam

** "The dead water extends up into the high land," Smith Brook

Allagaskwigamook begins. There the men would move up Sahbimskitegwek and scout all of the waters that fed that stream. They would also end up on the shores of Chemquasabamticook, where they would make contact with Running Bear if all went well.

"If the Malacite have a camp on Chemquasabamticook, then we will have a council with them so they know we will be hunting near them this season," said Running Bear. "If they have no camp there, I will send Hawk-In-Flight and Returning Thunder to Wallastook to find their main camp."

For the first time since he learned of the large burn at Pataweekongomoc, Running Bear felt somewhat at ease. It was due, he knew, to sharing the problem with others. Taking action, working together, agreeing on the best course of action—these were the Wabnaki ways. This was why the leader did not lead alone. Running Bear was reassured by the fact that the others felt the lands held much game, but even more glad that the others willingly shared the burden of decision-making.

Early the following morning, while the women and children broke camp and packed away the family's possessions, the men went to their canoes to pitch them. Pitching was almost a daily routine when the canoes were in constant use. The eyes of the birch opened up with some regularity and allowed water to seep in. Then, too, the accidental scraping and bumping of everyday use rubbed and chipped off the old pitch on seams and old repairs. These things happened no matter how carefully the canoes were handled.

Each canoe carried its own pitch pot, which was kept filled with hardened pitch. The pot, usually a hollowed-out stone, was placed near the fire and heated to melt the pitch. The pitch itself was actually an alloy made of spruce pitch, charcoal, and animal fat. The fat was the tempering agent, and the pitch was mixed according to the temperature of the season. During the summer months a much harder or brittle mix would be used than now when the water and air was much colder. The summer pitch would quickly crack and chip off during cold weather, and the cold weather pitch would soften and rub off if used during the summer season.

When there was a known leak in a canoe, it was marked to be pitched later. The minute holes that sometimes occurred in the eyes of the birch, however, were not always easily located. When a leak could not be seen at a glance, the repairman placed his lips over the suspected spot and by sucking in could determine whether there was a hole.

To repair a hole, the repairman first made sure the area was dry. If the sun was shining, it served as the best and easiest source of heat for drying. Without the sun, a firebrand was taken from the fire and applied to the broken area until it was dry. The melted resin was then applied to the bark with a stick and then smoothed and pressed into place with a moistened thumb.

With their canoes repaired, Walking Deer's band started loading while Running Bear and Moose-Comes-Running paddled out into A'pmoo jenegamook toward their camp to tell the family about their decision and to get them ready to join Walking Deer's band. Only a slight breeze was now blowing down the lake, and the air was so clear that they could see the distant shore as clearly as they could see the two small islands slipping by to their left.

Looking back down the lake Running Bear could see the great mountain Ktaadin, and because of the clear day it seemed much closer than it really was. "Inside the mountain lives Pamola," thought Running Bear. He knew that no one ever climbed the mountain, because if they did they would be eaten by Pamola. Still, his people's lives were tied to the mountain because they were tied to Gluskabe who came from there. Ktaadin was the center of their world.

It was mid-afternoon by the time both of the Bear Clan bands landed at Grey Wolf's camp. Grey Wolf and his family were so glad to see them that they would not hear of them leaving right away. Running Bear relented and decided he would spend the night rather that starting out for the Allagash stream immediately.

Grey Wolf said, "You will travel faster and safer with a good rest and a good meal." When Running Bear showed him the large supply of moose meat the clan was carrying, he said, "Now you surely must stay. We will prepare a great feast to celebrate our reunion."

Grey Wolf's wife, Singing Bird, agreed and immediately started preparations. The younger children were sent out to gather wood for the cooking fires. The children would later be given the responsibility for keeping the fires going while the meal cooked. The older girls went out and gathered roots and tubers along the banks of Nahmajimskit'egwek, an area which always yielded a generous supply.

Singing Bird went to her own supply of herbs and spices, which she used to give her cooking its special appeal. The two clans would feast on vast quantities of roasted moose meat, several varieties of vegetables, baked fish, nuts, and as a special treat, a pudding made of ground corn and sweetened with generous amounts of maple syrup.

While Grey Wolf's people made preparations for the feast, Running Bear and the men of his clan gathered bark for shelters while the women worked on building them. If they had been staying for a night or two, they could have crowded in with their friends, but as it could be several days or longer before the men returned, they decided to build lean-tos.

It was a wonderful feast and reunion. They all ate until they could hold no more. Then there was dancing, story-telling, and games for the children. The sound of their laughter rang out across the quiet waters of Pongokwahemook, but gradually the mood of the party grew more serious, and everyone drifted toward the central campfire. The children of both clans had been looking forward to one of Moose-Comes-Running's stories, and now they stared at him with expectant eyes shining in the firelight.

Moose-Comes-Running always enjoyed telling the children stories. Many tales he told were old ones handed down to him through countless generations of the Wabnaki people. Some of them, the less serious ones, he embellished, and sometimes he told stories of the deeds of his own clan so these too would become part of the Wabnaki heritage.

Tonight, Moose-Comes-Running put on his most serious expression, which the older children knew was only a mask, and said, "I am going to tell a story tonight that is a special story, but it is a sad story. It is a story of what can happen when children do not work hard and obey their parents. Do you all work hard and obey your parents?" he asked as his eyes swept over the group.

The children nodded and leaned forward, eagerly waiting for him to begin.

"A long time ago there was a family that used a certain place to hunt each season, just as we do. Like us, each spring after break-up they would pack and move back to their main village. Now, this was a large family. Seven children. And unless they all did their part, there would not be enough for everyone to eat. All in the family were hard workers except for the youngest, who was called Asa. Asa was very lazy. No matter what the boy's parents told him to do, he would not do it.

'Go tend the traps.'

'I'm too tired.'

'Get some wood.'

'I'm too tired.' Asa was always too tired.

"Finally, the father became disgusted with Asa. It was near spring break-up time, and the father decided to give the boy one more chance. He told him to get some wood, and if he did not do it he would be punished. Asa started out to fetch the wood, but soon became tired and sat down on a mossy bank, leaned back against a tree, and went to sleep. The father was furious. He woke the boy up and told him the break-up had started and they must break camp. 'Get up and help or we will leave you.'

'I'm too tired.'

"This was more than the father could take. The family set about breaking camp, and everyone helped: some loaded furs, others packed the bedding, some readied the canoes. Everyone had something to do except Asa, who slept on. When the family was ready the mother went to call him, but the father said, 'No, do not call him, he is no good. All he does is eat and sleep. I warned him. Now let him sleep.' They launched the canoes and started out for the village.

"The lazy boy slept on for three whole days, then awoke. He stretched and said, 'I'm hungry. I guess I had better go back to camp and get something to eat.'

"He got up and went back, but the camp was gone. He could not even find where it had been. He became frightened and did not know what to do or where to go. He was lost. He repented. He had been lazy and got himself into this situation. If he ever got out he would never be lazy again, he told himself.

"For many days the boy wandered about in the woods trying to find his way to the village. He grew more and more hungry, but there was no game to shoot, and he did not know what to do. At last he came to a small pond and gave up. He sat down on the bank and started to cry. Suddenly, he heard something and looked up to see a beautiful young woman standing on the other side of the pond. She said, 'Come here and cheer up.' Asa went to her, and she led him to her wigwam and gave him food.

"The girl was very beautiful and Asa was happy. Every day she would go into the woods and return with food, so they always had plenty to eat. After a time Asa fell in love with the woman, and they were married at the pond. Everything went along well for two years, and then the woman gave birth to a beautiful baby girl. Asa had never been so happy.

"They lived happily together for years, but finally Asa wanted to see his family and friends. He grew sad. His wife noticed this, but did not say anything for a long time. Finally she asked what was the matter. Asa said, 'Nothing,' but she knew better and told him he wanted to go back to his village, and he admitted she was right.

'Tomorrow morning take the canoe and go, but come back in seven days,' she said. He agreed.

"In the morning when he went to the canoe, he found it full of furs and food. He thanked his wife and left. Before he left, his wife told him where to go to find his people and to be sure to come back.

"Everyone at the village was surprised to see him. They thought he had surely died many years before. When they saw the furs and game, they were sure he had become a great hunter and were so glad to have him back that they gave a great party for him, exactly like the one we are having tonight. All the girls wanted to marry him. Everyone asked where he had been, but he would say nothing because he had promised not to. He paddled back to his wife, who was waiting for him.

"Asa and the beautiful woman continued to live happily together, but before long he wanted to return to his village. As before, his wife made him promise not to tell anyone about her and to return in seven days. Again the canoe was filled with furs and game and Asa left for the village. This time there was great rejoicing, and one of the girls tricked him into promising to marry her. So when he returned to his wife, he was very sad and didn't know what to do.

"His wife met him, and Asa told her what had happened. She said it was all right. She asked him to stay with her a few days before he returned to the village, and he did. When it was time to go back the canoe was filled again.

"Before he left, his wife said, 'You will never come back.' 'Promise me that you will never kill a white deer. In return you will be the greatest hunter in the whole village.'

"Asa promised, and then she said, 'When you are halfway home, look back and see us as we really are.' Asa said he would and paddled away."

The old story teller chose this moment to add to the suspense by rising to stretch for a moment. He found his hips and legs could not tolerate the long periods in one position as they had when he was younger. The children squirmed impatiently as he reseated himself and continued.

"Asa paddled away, and when he was halfway to the village he looked back. His wife and daughter were gone, and in their places stood a white doe and her white fawn. He raised his arm to them, and the two deer trotted away.

"Back at the village, the hunter married the girl who had tricked him, and they lived together in their wigwam. As he had been promised, he became a great hunter and his lodge was always filled with game. Always there was plenty of moose, beaver, bear, and furs, but his greedy wife always wanted more. No one in the village had as much, but still she wanted more. Then, too, she wanted him to tell her where he had been when he was away, but Asa would not tell. All of this time he never killed a white deer.

"The years went by and other hunters killed white deer, but not Asa. This angered his wife. She tried to make him kill a white deer, but he refused. Finally she made up her mind that he must kill a white deer and, when every trick she tried failed, she told stories about him. She said he was afraid to kill an albino and that he had evil power. She continued her stories until the village became angry with Asa for not killing a white deer. At last he agreed to kill one. His wife said, 'Never come back unless you bring a white deer.'

"Asa started out with some companions into a part of the woods where he had never been before. They came upon a white deer and his companions urged him to kill it, but Asa would not. The deer ran off, and the others became very angry with him. Finally, he saw another white doe, and this one stood still and would not run. He knew he had to kill it. He released his arrow and the deer fell dead on the snow. Asa ran up to it, and a voice said, 'You have killed me, father.'

"His companions came up and put the deer on a toboggan and started to haul it out of the woods, but they discovered Asa could not walk, so they put him on another toboggan and hauled him out too. As they traveled, Asa became worse and worse. By the time they neared the village, he could not move his arms and he felt numb all over. Meanwhile, the people of the village heard them approach and came out to greet them. Asa's wife was there, and she was so happy to see the deer that she paid no attention at all to her husband on the other toboggan, but set about to skin the deer immediately.

"Asa said, 'Do not skin it. Take it to my wigwam.' This made his wife furious, but there was nothing she could do, and the deer and man were hauled into the wigwam and placed by the fire. Asa was almost dead by now, but he told his wife to look and see what she had done. She refused because she was angry at the deer being brought inside. Asa insisted, and finally she was forced to turn and look. The dead doe changed as she looked at it into a beautiful young girl, shot through the heart. It was Asa's daughter, and as she changed Asa grew weaker and died. His wife watched, and as she watched she was turned into a great cat."

The children's expressions remained serious and somewhat sad, but it was late and the excitement of the day had taken its toll. They were ready for their sleeping robes.

[CHAPTER NINETEEN]

ROGUE BEAR

Instead of going back to the small stream where they had entered Pongokwahemook, Running Bear and the others paddled along the southern shore until they reached a point where only a narrow piece of land separated the two lakes. Here they carried across into A'pmoojenegamook, and at that point were only a short distance from where Allagash Stream enters the lake. This shallow part of A'pmoojenegamook was a favorite feeding ground for moose during the late spring and summer. Now, as the leaves on the trees were beginning to turn brown and fall to the ground, the great moose were seldom seen feeding in open places. During the fall and winter they would be found feeding in the higher country, eating the browse of both evergreens and hardwoods.

The two canoes entered Allagash Stream, and the men soon put away their paddles and took up their poles. They were traveling light, only the barest necessities, so the canoes skimmed easily over the water even though there was a good flow running against them. Expertly the two men in each canoe stood and used their poles to work the canoes upstream. While one held the canoe in position against the current, the other found new purchase with his pole and sent the canoe forward by climbing the pole hand over hand. Then he held while his partner repeated the process in the other end of the canoe.

The heaviest person stood in the rear of the canoe along with the gear. So the back of the canoe was lower in the water than the front, and the current tended to help head the craft upstream. When poling downstream they swapped positions so the front of the canoe rode lower in the water.

By noon the men reached the waterfall and the small pond that marked the halfway point to Allagash Lake. In one trip they had the canoes and their belongings up to the pond above the falls and were ready to continue their journey. They paused for a while, resting in the sun on the ledges above the falls.

The stream above the falls was even swifter than below, but only twice was it necessary for them to get out and carry their canoes, and then only a short distance to get over ledges that created a small waterfall. If the water had been lower it would have been necessary to get out of the canoes in some places and wade, towing them behind to avoid scraping on the rocks. But with the water level adequate, the canoes floated just fine in the hands of the expert canoemen. They seemed to skid over the fast-flowing water like leaves in the wind.

By late afternoon they reached the lake. As they emerged from the dense cover of the trees they could feel a fresh, cool breeze blowing from the north. "It will not be long before we will be trading our canoes for snowshoes," Moose-Comes-Running said. "That wind reminds me that winter is not far away and the first snows will soon be here."

"Yes," Running Bear agreed, "I wish to complete this trip as rapidly as possible. Let us hope the cold does not come too fast for us to use our canoes on the beaver ponds as we scout this new area."

They paddled across the north end of the lake and made camp on the south shore of the cove, where Allagash Stream entered the lake. On the hill behind their camp, caves went deep into the ground. The walls of these caves were always wet and covered with a silvery substance that looked like ice. They were called the ice caves, and ice could sometimes be found there during the hot summer months. Hawk-In-Flight had never been to the caves before, so he and Returning Thunder went down to explore them while Running Bear and Moose-Comes-Running discussed the scouting ahead of them.

Here the party would separate, Running Bear and his son in one canoe and Moose-Comes-Running and Returning Thunder in the other. This way they would complete the scouting twice as fast and be able to return to the family on Pongokwahemook. Plans were made to rendezvous at Chemquasabamticook if convenient, but neither pair would wait there unduly long for the other, for it was more important to complete the survey and return quickly. If they could meet, fine. If not, no harm would be done. They would all regroup at Grey Wolf's camp.

The days that followed were joyful for Running Bear and his son. The need to scout new hunting grounds made it necessary to do what they enjoyed most—travel and observe new country. They carried their canoe from beaver flowage to beaver flowage. At times they left it to explore small streams and the high ground around them on foot. The more they traveled, the more elated Running Bear became. They were finding more game than he had dared hope for. He knew now that this area would more than make up for their loss in the Pataweekongo'moc area.

The area they were in was the headwaters of the whole Allagash Waterway. Somewhere here was a small stream, higher than all the rest, that was the headwaters of the river called Allagash.

In due time father and son made their way to Chemquasabamticook, where they made camp on the east shore. Crossing the lake, they found totems of one of the Malacite clans and an abandoned hunting camp. Rather than try to make contact now as he had originally planned, Running Bear decided to leave a message and visit the camp later during winter. Back on the eastern shore he left his totem and a message for Moose-Comes-Running and Returning Thunder. Early tomorrow they would start back for Grey Wolf's camp. During the winter hunt, the clan would post the boundaries of the new area and establish trails that would connect with the network already in place in the old hunting grounds.

When the pair left Chemquasabamticook they carried the canoe directly south to one of the branches of Sahbimskitegwek, which would eventually lead them to Pongokwahemook. Even the smallest branches of the stream had beaver in them. The canoe traveled fast. They only needed to lift it over the dams and sometimes for a short distance below before they came to another flowage. They traveled rapidly through this area because they knew Walking Deer would have scouted the area already. They never met the others, but this did not concern Running Bear beause he knew the country was vast. Besides, there was really no need to meet.

After two day's travel they joined their family at Grey Wolf's camp, and on the following day the rest of the scouting party returned.

The women had not been idle while the men were away. Work had started on the winter garments that would be needed to keep them warm through the long winter.

Hawk-In-Flight's wife, Autumn Leaf, could not wait to show her husband what she was making for him—moose-hock boots. She had peeled the skin from the hind legs of the moose he had killed down to the middle of the lower joint. Here she cut through the hide. She would sew the pieces tightly after they were cured. The hair of the moose would make for a warm, nearly waterproof, pair of boots. The moose's hock would form the heel.

Men's clothing consisted of moccasins, leggings, shirt, and breechcloth. Women wore moccasins, leggings, and dresses. Women also fashioned coats from furs for winter wear. The arms of these coats were removable, an important feature allowing hunters freedom of movement and permitting them to cool off while traveling or under exertion.

Sugar Maple was skilled at making these garments, and she took pleasure in decorating them when time allowed. She used dyes to color porcupine quills, which she worked into intricate designs. Sometimes she used the dyed white hair of the moose to embroider, and she often worked wampum beads and feathers into the patterns. One of her favorite designs was a double curve, said to have been a gift from Gluskabe.

Many variations of this design were used, and it was applied to canoes and tools as well as clothing. The women had also laced several pairs of snowshoes, and some of these were traded with Grey Wolf for stone arrow points and other fine tools that he was skilled at making.

Grey Wolf's sons and grandsons also worked with stone. The younger ones started out by spending seemingly endless hours polishing cutting edges with another stone. Somehow, though, after the long periods of boredom they discovered the pride that comes with good workmanship and continued to learn the craft from the master. Some of them now could chip a point nearly as well as Grey Wolf. He proudly showed his friend some of these tools. Running Bear was glad to trade for these items. He had learned the rudiments of the craft, but he never cared for it.

Walking Deer was as enthusiastic about the area he scouted as Running Bear was about his. Walking Deer had camped one night with a party of Malacite hunters, so that tribe was now aware of the Bear Clan's new hunting area. The Malacites also had a story they knew would be of great interest to Bear Clansmen.

One night, about a week earlier, the Malacite hunting party had awakened to the sound of a ripping and tearing at the side of their wigwam. The men immediately rose and chased a bear away by yelling and shouting. They did not want to attack the bear in the dark for fear of wounding it. When the hunters were satisfied that the bear was gone, they returned to their robes to go back to sleep. Being somewhat nervous about the whole affair, they were still awake when the bear returned. Again, they chased the bear away without seeing it, but this time they decided they had better build a large fire and keep someone on guard. For the rest of the night, the bear could be heard circling just beyond the light of the fire. No one slept.

At dawn the hunters decided they had better attack the bear before it attacked them. It was obviously very hungry and could smell the food they had with them. The bear was becoming more and more bold, and now that the men could see it they knew it was a sick bear. It was very thin and probably very old. Armed with bows and arrows, they started to shoot at it. Two arrows struck, and the bear roared like an angry man and swatted at the arrows with its huge paw. It left limping and was not seen again. The hunters didn't think their arrows wounded the bear, but rather that he was limping because of an old injury.

The Malacites trailed the bear to a beaver flowage, but they could not find where it emerged from the water. They did not think the bear was weak enough to have drowned, however. They were sure the bear was too thin to go

into hibernation this winter and would become meaner and more dangerous as food became scarcer.

Running Bear and the others were disturbed by this. Awe'sus, the bear, was their totem. "Totem" meant the same thing to all Wabnaki tribes and other tribes to the south and west: "My relative" or "My special friend". This did not mean that the clan never hunted or killed a bear, but only that the bear was of special significance to them. They avoided killing female bears and never allowed a dog to chew a bone from a bear, but to have a taboo against killing all bears would be as unthinkable as a taboo against killing all other animals. Were not the other animals their brothers too? The animals must die that The People may live. These facts were basic. In any case, the bear had to be destroyed. The bear's dignity demanded it, and so did the welfare of the people who might encounter it. Running Bear vowed to hunt down this bear. It was the clan's duty.

Walking Deer's band was to winter in one of the traditional camps on Allagaskwigamook. Running Bear would establish a new camp closer to the hunting grounds. Walking Deer's camp was about a third of the way up Allagaskwigamook, near a spit of land that ran out into the lake. The Bear Clan had used this site and others near the lake for generations. In fact, the lake was named after these hunting camps.

The site Running Bear had in mind for his camp was on the north end of Pongokwahemook, where the lake narrowed and became a thoroughfare that connected to Lake Pataquongamis, which was joined to Lake Allagaskwigamook by another thoroughfare.

In the morning the Bear Clan packed and headed for its permanent hunting camp. Running Bear had not selected the exact spot, but he knew the area well. As he approached the narrows, he looked at the west shore and knew that was where they would build their camp. The area had plenty of tall firs and spruces to protect them from the cold north wind. It offered a good view down the lake, and the two bands of the clan would not be too far apart. This last point was important in the event of a food shortage. The clan must be close enough to share food, yet, far enough apart to take advantage of all the game available, large and small.

Unlike Walking Deer, who would be moving into a camp previously used by the Bear Clan, Running Bear's band would build a new camp from the ground up. This was no serious disadvantage, however, because they knew they could build new winter quarters almost as fast as older ones could be repaired.

As the family went ashore, Walking Deer and his band continued on the thoroughfare toward their camps. Farewells were brief because they all knew they would see each other often during the long winter.

While the family unloaded the canoes, Running Bear, Moose-Comes-Running, Returning Thunder, and Hawk-In-Flight went up the bank to find a level area to erect their wigwams and establish the new camp. Then Running Bear, Hawk-In-Flight, and Returning Thunder marked a circle on the ground where each would set up his wigwam. The dwellings would be arranged so the entrances faced south and all opened into a common area. The circles for the wigwams were about four paces across, but Running Bear's would be slightly larger than the others because Moose-Comes-Running shared it with him and his family.

The families cut and gathered saplings, which would be placed at intervals around the circle and joined at the top, forming a cone of slender trees. The saplings had to be set into the ground so the structure would be sturdy enough to withstand cold windy days ahead. As Running Bear and Moose-Comes-Running started scraping out these holes with a flat stone and a stick, a familiar shape caught Running Bear's eye. He picked up a perfectly shaped arrow point made from the same stone that Grey Wolf obtained from Lake Munsungun. He showed it to Moose-Comes-Running, and both sat back on their heels to reflect on the find.

Running Bear said, "It is the same material that Grey Wolf uses, but it is shaped different than our arrow points."

Moose-Comes-Running ran his fingers through the loose dirt and picked up chips of the same material. "Look," he said, "here's more. The point could have been made here."

The two men talked for a time about what they had found. No one remembered ever using this place to camp, nor did they remember their father's father speaking of anyone who had used this place. The point was shaped differently. Did this mean that another people made these arrows and dropped them here, or did it mean that their ancestors made their points slightly different from theirs?

They both knew they couldn't answer these questions, at least not now, but they would save the point and show it to Grey Wolf. One thing was sure. The Bear Clan was not the first to camp on this spot. Sometime, perhaps many, many seasons ago, Running Bear thought, another clan leader had been drawn to this exact spot by the same combination of things that had made it attractive to him.

Once the cone of saplings was in place, it was covered with mats of birch bark. Some of these were carried with the families for temporary shelters along the way, but more would be needed to make the permanent camp warm.

The mats were made of birch bark strips, an arm's length or more long and of various widths, sewn together along the widths, slightly overlapped. Split spruce roots were used as thread. Two narrow strips of wood were attached to the ends. They were easily rolled and transported when necessary. The mats were tied to the sapling frame and overlapped to keep out the rain. The low entrance was covered with moose or caribou hide to keep out the wind. Hides and furs were hung inside the shelter to provide insulation.

The wigwam was heated by a central fire. The smoke escaped through an opening at the top of the structure. In extremely cold weather a large fire was built outside in the central area, and each family placed large stones close to it. When the rocks were hot they were carried into the wigwam between two sticks to make the shelter warmer.

Before darkness fell the three wigwams were finished, at least the outer structure. Fireplaces were built in front of each, and the women started cooking the family's supper. In the days to come everyone would have to work hard to make the camp snug and safe for winter. But for now the families were home, as content as they would be in the permanent village on the Penobscot. The security of being with the family and clan was the important thing to each of them, not where they lived or how long they stayed there.

In the days that followed, Running Bear and the band were extremely busy. There was much work for everyone, and as the days grew shorter there was a frenzy of activity to ensure the clan would be ready for the arrival of pabo'ni, winter.

The women tanned hides for robes and clothing, sewed, and made moccasins. And there were showshoes to lace, bark containers to make, and pemmican to prepare.

The pemmican was the hunter's trail food. It was made of equal portions of lean meat and fat. The meat could be moose, caribou, or deer. Any kind of fat could be used, but bear fat was preferred. The meat was first cut into strips and dried. Then it was pounded into flakes before it was mixed with melted fat. The pemmican was mixed, stored, and carried in the skins of small animals that were especially prepared for this purpose. When available, dried berries were added to the mixture, and if there were any maple syrup or sugar left it too could be added to make the mix all the more palatable. Honey was also used as a sweetener when available.

Each of the men was busy establishing his trap lines and making mental notes of where the most game sign was located. They had covered as much of their hunting area as possible. Boundaries were marked, trails established, and all was made ready for freeze-up. Travel was limited now. The big lakes were open so the canoes could be used there, but the small marshes and beaver flowages were covered with ice. Canoes could not be used on them, but neither was the ice strong enough to hold up a man. All they could do was wait for the cold that would make the ice strong and bring the snow.

[CHAPTER TWENTY]

THE GOOSE HUNT

While the lake was still open, Running Bear and Moose-Comes-Running took the opportunity to paddle up to Walking Deer's camp. They had the rogue bear still on their minds and wanted to discuss the problem with Walking Deer.

The men sat and smoked and talked about the bear and what could be done about it. There was not much they could do, they decided, until someone saw the bear again or, when the snows came, saw his tracks. When this happened, whichever camp spotted the animal or its tracks would send a young man to the other's camp and a hunt would quickly be organized. Talk of bears fascinated the young boys who sat nearby listening to the men talk. During a lull in the conversation, a small boy called Little Otter summoned the courage to ask, "How did the Bear Clan come to select the bear as its totem?"

Though Moose-Comes-Running was the recognized story teller of the clan, he deferred to Running Bear in this case, because as head of the clan it was his place to tell the story of how the Bear Clan came to be.

"It really happened the other way around, Little Otter," he said. "Many seasons ago a man and a woman and their little boy were traveling from one camp to another. When the family came to a carry, the father took the canoe and headed for the other side. The mother picked up a load and did the same. Each parent thought that the little boy was with the other, but he was not and wandered off in the woods. It was a very long portage, so it was a long time before the man and woman missed the little boy. When they realized that the boy was gone, they hurried back over the trail, but their son was nowhere to be found. They searched for days and days through the woods, but they saw no sign of the little boy. Finally they had to give up and go on. They were very sad.

"The boy wandered around and around in the woods. He could not find his mother or father or even the trail they had taken. For many days he wandered, and he grew very hungry and very tired. He was afraid, and he knew he would never see his mother or father again. Finally the boy lay down by a small pond. He was too tired and hungry to go on. He lay there for a long time crying softly, and when he opened his eyes a big sow bear was standing over him.

"The bear had a cub with her, but her other cub had just been killed by a wolf, so she had plenty of milk. She let the boy drink the milk, and soon the boy was sitting up and feeling much better. The bear took the boy to her den, and she raised him along with the other cub.

"As the boy grew into a young man he realized that he was different from his brother and his mother, and he asked her about this. The bear told the boy how she had found him and that now she would take him and show him his own kind. The young man was not sure he wanted to do this, but he went with the bear to the edge of one of the villages of The People.

"The young man was so fascinated at seeing his own kind for the first time that he walked right into the village without even realizing what he was doing. No one in the village knew the young man, but he was made welcome. When he turned to look for the mother bear, she was gone, so he decided to stay with The People.

"The man proved to be a great hunter. After a while, he met a beautiful girl and they were married. The young man was very happy with his new wife, but he was also very sad that his mother bear could not share his happiness. One day as he and his wife were walking in the woods, the bear appeared before them. 'You are Awe'sus, a bear,' she told the man. Then she turned to his wife and said, 'In marrying Awe'sus you too have become a bear, and so shall be your children and your children's children.' Then the bear turned and was gone.

"The young man's wife was very frightened by this incident, because he had never told her about his life. But now, as he told her about it, she wished the bear had stayed so she could thank her. Instead, she and the man promised each other they would remain worthy of the name Awe'sus and see to it that their children were taught to do the same."

Running Bear got to his feet and said, "So you see, Little Otter, we did not select the bear as our totem, but instead we were selected by the bear. "

While the story he had heard so often was being told, Walking Deer's mind was on other things. He had been thinking about the great flocks of geese and ducks he had seen heading toward Umsaskis Marsh on the north end of Umsaskis Lake. Finally he said, "The river between Allagaskwigamook and Umsaskis has enough water to float our canoes if their load is light, if we use our poles, and if we're careful. I've seen many geese and ducks heading into and out of Umsaskis. I think we should go down to hunt them."

Running Bear and the others perked up. The relative inactivity lately was making them restless, and a goose hunt sounded most appealing.

"Even if the marsh and lakes freeze over while we're gone, we could still pole our canoes up the river as far as Allagaskwigamook, " Walking Deer added. "Then, if necessary, we could leave them there and return to our camps on foot."

The others quickly agreed, and Running Bear and Moose-Comes-Running returned to their camp to prepare for the hunt.

Early next morning Running Bear, Moose-Comes-Running, and Hawk-In-Flight paddled through the thoroughfare, through Pataquongamis and into Allagaskwigamook. This time Returning Thunder had been chosen to remain at camp. The morning was cold, and there was ice near the shore when they left, but the sky was clear and crisp. High overhead they could hear the honking of the great geese. Looking up, the men saw the birds heading south in their perfect "V" formation. As they watched the flocks against the deep blue sky, each man felt a deep reverence for this great bird that each fall headed south and then returned from that direction each spring. No one knew where they went, but it was known that many of them raised their families in Umsaskis Marsh and many more stopped over at the marsh on their way south.

"Sometimes, when many flocks are seen heading south, it means a storm or cold is coming," Moose-Comes-Running said.

"If so, let us hope it waits until we return to our camps," Running Bear replied.

By this time Walking Deer's group had seen them and paddled out to join them as they continued up the lake to where the Allagash River left Allagaskwigamook.

Allagash was a name the people used when they spoke of the river in general terms. It was a shortened version of Allagaskwigamook. It was not the Wabnaki way to name a whole river. They did not think of it in that way. Instead they named specific places along the river's course, and these were the names most often used in conversation. A section might have been named after a natural feature. For instance, "Umsaskis" means "place having opposite points which run out to meet one another, " and the name eventually came to mean the whole lake. A place might also have been named for a happening. If someone lost a prized axe in an area, the place might have become known as: "The place Standing Bear lost his tomhe'gan (axe)." In time the name might have been shortened to "Tomhe'gan Place" or "Tomhe'gan. " (There is a place named Tomhe'gan near the great Xsebem. The axe was lost by Running Bear's father, Standing Bear.)

Soon the group was passing through a narrows and into an area of the lake known as "Place of the heron." During the summer when they fished in the shallows, they saw many of these long-legged birds. Today, the men saw no herons, but they could hear the roar of the river as it ran over the rocks on its way to Umsaskis.

The party went ashore above the rapids and walked down along the shore to see what they were up against. They were tempted to try the run through the rapids, for they all greatly enjoyed the wild rides through white water and great standing waves. But this time they had to be practical. The water was low, and there were places where it would be extremely difficult to avoid rocks. No one relished the thought of a dunking in ice-cold water. So they decided to carry the canoes over the first and steepest drop of the river and then pole the rest of the way down.

The carry trail around the rapids was well-worn. It had been used for generations. Sometimes only short sections were used. But in the spring when the water was high and roared down the entire length of the rapids, travelers moving upstream would have no choice but to carry along the entire trail, because it was impossible to pole against the powerful current. A skillful paddler might ride the spring flood downstream, but he would do so knowing that he risked his canoe, perhaps even his life.

But with the water low, Running Bear's party had little trouble poling up and down the rapids. The men stood in their canoes and gently eased the birch vessels around rocks, through chutes, and over quiet sections of water. Occasionally, where the water became shallow, they had to step out so the canoes would not scrape bottom, but these places were only ankle-deep, and it was often possible to step on rocks to avoid getting wet.

As the hunting party passed some islands they saw a break in the trees to the left and knew they were approaching the marsh. Just downstream the river turned left and became deeper. The canoeists knelt and gripped their paddles. The current was swift, and the men were soon climbing a high bank on the right shore where a small brook entered the river.

They could see some of the marsh from here. There was ice in some of the small quiet channels, but most of the larger ones were still open. They decided to travel to the south shore of Umsaskis, where there were shelters they could use. As they stood on the bank they could hear the sound of geese talking to each other, and the sound made them all the more eager to get on with the hunt. They would go to the camp, leave their gear, then paddle into the marsh for an evening's hunt.

The night and the day that followed were cold. The unfrozen areas of the marsh became smaller and smaller. The weather helped the hunters because it forced the remaining geese into the small areas of open water to search for food.

The hunt went well. To find the areas the geese preferred, the men paddled to the open areas and looked for feathers floating on the surface. When they found a likely spot, one of the men would hide out and wait for a flock of migrating birds to land or some of the resident geese to swim within range. It was not an easy way to hunt. The long periods of sitting or lying in the cold and dampness caused sore muscles and stiff joints. The hunters had to remain motionless until the geese were close. No bird was more wary than the great white-cheeked geese, so one careless moment on the part of the hunter could mean hours of patient waiting wasted.

When the flock of geese were close, the hunter shot the nearest bird and then quickly fit another arrow to try for another before the geese swam out of range. The birds were large and heavy, so once they were in the water they were slow taking off again. Once airborne they flew very fast. Sometimes, if the bowman was unusually quick it was possible to get three geese. Running Bear and some of the older hunters had accomplished this feat, and Walking Deer succeeded in doing it on this hunt. As this was one of the marks of a very skillful hunter, the story was recounted in vivid detail as the men sat round their fire that night. Everyone listened intently as Walking Deer related his experience, especially the younger hunters who had not yet reached this level of skill.

All through the day the hunters had seen many, many geese. Most were flying overhead, but a few stopped at the marsh. The wind had been shifting around all day, and the sky had become overcast by evening.

"I think we are going to see a storm," said Moose-Comes-Running. "It stays cold so it will be snow. I think pabotni, winter, is going to arrive soon."

"We will leave here to return to our camps at first light in the morning," Running Bear said. "If we hurry we can be back before another night falls and before the storm starts."

By the time they crawled into their robes that night, the wind was starting to pick up and could be heard whispering through the trees above them. They were in a sheltered spot, but during the night they were awakened by the rattle of snow on their birch bark lean-to and the howling wind from the northeast.

"I think we will be here longer than planned," Walking Deer said. Then, because there was nothing they could do, they all went back to sleep.

When daylight came, the snow was still falling and the wind was howling out of the northeast. The snow was piling up around their lean-to where there was shelter from the wind. But out on the lake, driven by the wind, the flakes appeared to be moving horizontally instead of falling to earth. On the lake great waves rolled by, and the wind whipped the tops of them into a white foam. The men were relatively comfortable in their lean-to, but knowing the unpredictability of storms such as this, they decided they had better prepare for the worst.

They had plenty of food, so that would not be a problem. However, if the snow continued to pile up they would have more and more difficulty moving around without snowshoes. So they gathered a large wood supply. The snow was nearly to their knees in open places. When the wood was piled up near the lean-to, they cut fir boughs and used them to close in the front of the shelter. When they were finished, they had a cozy wigwam made up of the birch bark that was the original lean-to. The boughs soon filled with snow to form a solid, draft-proof front wall. There was an opening at the top for smoke to escape, and after making sure that the canoes were in a sheltered place and out of harm's way, they settled down around their small fire to wait out the storm. If it didn't last too long, they would enjoy the storm-imposed period of inactivity.

Moose-Comes-Running said, "The lakes will not freeze over while the wind blows, so even though pabotni is here, we will still be able to use our canoes to return to our camps when the storm is over. I am glad we are here and not on our way to the camps."

[CHAPTER TWENTY-ONE]

RUNNING BEAR'S JOURNEY

Although none of them would say so, each of the men sort of enjoyed being storm-bound in this cozy place. Their families were safe in their winter camps with an adequate supply of fresh food, so they did not have to worry about them. They themselves were safe and comfortable with nothing to do but sit around and tell stories and recall past hunts. Hawk-In-Flight, who had never been farther north on this river than where they were camped now, asked his father about the rest of the waterway. Running Bear packed his stone pipe, picked up a coal from the fire to light it, and began.

"One winter, many seasons ago when I was a very young man, The People had a very hard winter in the hunting grounds. The snows were deep, and the bitter cold was with us for a long, long time. The animals were scarce and would not allow themselves to be killed, and there was nothing to eat. The hunters went out, but being weak from hunger, they could not travel far. We gathered bark from the trees, and the inside of the bark was scraped off. This we chewed, and it was all we had to eat for many days. We also chewed the leather laces, and even our moccasins to help ease the pain in our stomachs.

"Many of our people did not live until spring to go back to our village on the Penobscot. My grandfather died, and one of my younger sisters was not strong enough to live until si'gwan, spring, came.

"The Malacites from Wallastook were also starving that winter. One day two young Malacite men staggered into our camp. They had been hunting and found themselves so far from their camp that they did not have the strength to return. We had no food to give them, but we gave them some bark to chew and made them welcome. The two stayed with us until spring break-up, and then when it was time for them to leave it was very bad traveling for them to go through the deep, wet snow across country to Wallastook.

"My father said he could not spare a canoe for them to take, but asked me if I wished to take a canoe with Grey Wolf and the two Malacites and travel down the Allagash into Wallastook, then down the Wallastook to the village of the Malacite. We would leave the young men at their village, and then Grey

Wolf and I would continue down Wallastook to the great salt water. Once there we would follow the coast to our tribe's summer village at the mouth of the Penobscot.

"Of course, Grey Wolf and I were eager for such an adventure, and after a few weeks of hunting and feasting on the returning birds and animals, we headed north up Allagaskwigamook. Meanwhile, the Bear and Wolf clans paddled toward Pongokwahemook, starting their move back to our main village and eventually on to the village near the great salt water.

"With our bodies strong again with spring food in our bellies, the four of us in one large canoe were eager to get started on our journey. The Malacites were about the same age as us, and we had become good friends. We arrived at the rapids where the river leaves Allagaskwigamook by mid-morning. My father had cautioned us to use great care at the rapids and said it would be best to carry the canoe around the first drop. But we were young and had more confidence in our abilities than we had wisdom. So we headed the canoe into the roaring rapids. It was not long before we were moving so fast that it became impossible for us to avoid hitting a rock. When we did, the water came in and made our canoe so heavy that we quickly hit another rock. At this we all leaped into the water and, being close to shore, managed to get the canoe out of the current before there was more damage. We were fortunate that the damage was repairable, and by late afternoon we continued, with much caution, through the rapids and finally out onto Umsaskis. We spent our first night at this very camp. This is how I came to travel down this river. Now I will tell you what we saw."

Hawk-in-Flight went outside and brought in more wood and added a few sticks to the fire. Then he settled himself to hear the rest of the adventure.

Running Bear continued, "We paddled down Umsaskis, and on the right side where there was a point running out into the lake with large rocks and ledges on it, we saw an old campsite. The Malacites said that sometimes one of their tribe's clans came up the river this far to hunt. When we reached the end of Umsaskis the lake narrowed to a thoroughfare and, though the water appeared to be still, it flowed with a strong current. Very quickly we found ourselves entering a very long, narrow lake, and the Malacites said this was the regular hunting grounds for some clans of their tribe. They pointed out several now-abandoned camps, and there were moose hide-drying frames and wigwam frames in several locations as we traveled down the lake. At the north end of the lake we saw a large meadow; it was the place where the stream that flows from Chemquasabamticook enters this river. The Malacites told us this was a place of much game, especially moose.

"We paddled by the meadow formed by the stream from Chemquasabamticook, and with some current soon entered another small pond, which narrowed and became river again. There were also hunting camps along these shores and even along the river as we left the small pond. The Malacites said some of their best workers of stone worked at these river camps, and they sometimes left unneeded materials hidden here until they returned in another season. Grey Wolf was greatly interested in the work of stone-cutters, but as there was nothing to be learned from stopping, we continued.

"We found the river below these lakes to be fast-flowing but with no dangerous rapids that would damage our canoe. The river flowed north for a while and then, by the position of the sun, we could see that we were moving almost south. The Malacites said that this meant we were not very far from another Pataquongamis, and indeed we soon found ourselves in an area of islands and meadow. Each of the islands was covered with curled up young fern* that is so sweet in the early spring. We picked a supply of them and then continued to the pond. There were hunting camps on this pond too, but we only saw one because the outlet is very close to where we entered, so it was not necessary to paddle the length of the pond.

"We stayed at a camp between the inlet and the outlet for the night. That evening we went fishing among the islands where the river enters the pond. We had hooks made from the breastbone of the partridge, and we tied on deer hair as a lure. We caught many trout that evening, and what a feed we had that night with the freshly caught fish and the freshly picked greens! I remember thinking that I could imagine no place I would rather be than right there on Pataquongamis with these friends.

"We enjoyed that place so much that we decided to stay another day. We fished, explored, and picked more greens. Our snares caught rabbits that were now growing fat from the fresh food that was starting to grow everywhere. We saw deer on the shore of Pataquongamis, but we did not hunt them, because we had plenty of food. On the shore, between the inlet and the outlet, we found a spring where we refreshed ourselves with the cold, clear water. The winter of starvation that had been so hard on The People now seemed far, far away to us four young men out on a carefree adventure together.

"When we left Pataquongamis, the river was a little steeper than it had been above the pond, but the water was still high with the melting snows, and we carefully made our way down with poles and paddles until we reached a place where the river turned east. Here the water seemed to stop, although when we looked at the weeds on the bottom we could see that the water was still flowing with a strong current. A stream called Musquacook enters the river in this deadwater.

"The Malacites told us that this was their canoe builder's favored area for birch bark. Of course, I was interested in this, because I was learning

* Fiddlehead fern

the craft from my father. However, the best time for gathering canoe bark had just passed, so, though I saw many good trees, we continued.

"The river flowed quietly for a long time and then, at a place where a brook entered the river like five fingers of your hand, the current started to move fast again. The character of the land was starting to change. We could see more hills, and very soon we entered an area where the river separated into several channels. We again found ourselves among many islands. We stopped and picked more of the curled-up greens, and we asked the Malacites if these islands meant we were approaching another pond. They laughed and said no. They said that the opposite was true, that we would soon see the highest falls on all of the Allagash and Wallastook waters. They said they were not sure about the Penobscot, but they doubted if any there were as high.

"As we approached the falls we came to a turn in the river where the water was quiet, but when we rounded the corner we could hear the great roar of the falls and that of the rapids that preceeded them. We quickly went ashore, and the Malacites led us to the top of the falls, where Grey Wolf and I stood in awe of the thundering water. It was not just the sight of the snow-white, churning water as it dropped two canoe lengths, but also the sound that went with it. Grey Wolf and I had to shout into each other's ears to make ourselves heard.

"Looking down below the falls, we could see men and women along the shores. They were fishing for salmon, just as our people do at the falls along

Painting by Louise Gilpatrick Rice

the Penobscot. The two Malacite men who were with us had hurried down to where they were because they had known their people would be here. None of the young men's clan were there, but everyone was very happy to see them. The young hunters had been given up for dead by their families.

"We spent the night with the Malacites, and it was the first time since leaving our families back at Allagaskwigamook that we had a meal prepared by a woman. As we saw how happy the people were to have their two young men back, we realized how much we were starting to miss our own people. It was a happy time. There was plenty of the great silver fish to eat, and the women added their own herbs and spices that made everything taste so good. We ate until we could eat no more, and then we listened to the stories of the Malacite people. Grey Wolf and I came to realize how close to being one all of the Wabnaki were as we listened to familiar stories and the adventures of familiar heroes. Though we knew that Gluskabe was not the exclusive property of the Penobscots, this was the first time we had heard another tribe talk of him and tell stories about him

"When we went to our robes that night we were both deep in thought. Though we did not know it then, we were a lot closer to manhood that night than we had been that morning. We both lay awake for a long time listening to the roar of the falls and thinking about what had happened that day. When I thought back on this trip in later years, I knew that this day had proven my father's wisdom in sending Grey Wolf and me on this journey.

"Grey Wolf and I would have liked to spend another day at the great falls of the Allagash, but the two Malacites were most anxious to get to their village and see their families who mourned them as dead. So early the next morning we carried our canoe and equipment over the large rocks that were the carry trail, and we were on our way downriver again.

"The countryside and the river took on yet another change. Now we were surrounded by rolling hills. The river surface seemed to be that of a long lake tilted so we seemed to be on flat water much of the time but it appeared to be on the side of a hill. This made a good current, and we were soon at the great twist in the river, which the Malacites said meant we were about to enter the Wallastook. After heading south and then turning north again we entered Wallastook, where there were several islands.

"Wallastook flowed east here, and the river was even larger than our Penobscot. A deer on the opposite bank was like a grain of sand, and had it not moved we would not have seen it at all. We went through several rapids that day on the way to the Malacite village, but the water was still high, and we had no difficulty getting through without damage. By late afternoon we approached the Malacite village, which was located on the north bank where another river* enters Wallastook. The people saw us coming, and before we

* St. Francis River

were ashore they had recognized the two young men with us and were yelling for their families. Everyone was so happy to have the two young hunters back that Grey Wolf and I started to wonder if we had been noticed at all. But soon the attention turned to us and we were showered with thanks for bringing the young men back. We said that we had really done nothing to deserve thanks, but they would not hear us and set about preparing a great feast in our honor.

"We did not know it until then, but the father of one of the young men was the Sahgum* of the Malacites. After we feasted that night, he spoke at great length about how the Penobscots and the Malacites were all brothers in the Wabnaki confederation. He spoke of how no tribes of the Wabnaki had power over any other tribe, but the independant tribes were bound together for mutual benefit and protection. He named some of the tribes: the Norridgewocks, Sagadahocks, Penobscots, Passamaquoddy, Malacite, and Micmacs. He said he had attended many councils of these and other tribes. At each council a different Sahgum acted as leader, and as a result of this alliance there had never been a war among the Wabnaki in anyone's memory.

"The Sahgum told how, unlike our enemies the Mohawks, the Wabnaki honored their women and allowed them equal voice in council. In some tribes of the Wabnaki some Sahgums were women. Before the Sahgum ended his speech he gave Grey Wolf and me a wide belt of finely made wampum, which was to be carried back to our Sahgum in thanks for returning their young men. Grey Wolf and I accepted the belt, and we both felt very proud to be a Penobscot and a Wabnaki."

Running Bear stood and stretched. He had been sitting for a long time. It was dark outside, and the storm continued to blow. The hunters ate some roasted goose. "Grey Wolf and I remained with the Malacites for seven days before continuing our journey," Running Bear said, "But I will have to tell the rest of the story some other time. It is late, and besides, I have answered Hawk-In-Flight's question about the rest of the Allagash. "

* Chief

[CHAPTER TWENTY-TWO]

WINTER CAMP

When the men crawled into their robes, the wind continued to howl through the trees. The sound of it lulled them to sleep. Sometime during the night Running Bear awoke and for a moment could not figure out what had awakened him. Then he realized it was quiet! The wind was not blowing. The storm was over! In the morning they would head back to their camps and their families.

At dawn the hunters pushed their way through the deep snow and dug out their canoes. After a quick meal of cold goose meat, they loaded them and were on their way back up the river before the sun was above the southeastern horizon. It was fortunate that the lakes were not frozen, because otherwise they would have had to travel on foot, and the deep snow would have made it difficult and very slow. All of them knew how to make temporary snowshoes from saplings and hardwood twigs, but these emergency snowshoes were cumbersome and in no way equal to the finely crafted ones that Running Bear made.

The day was cold, but as the sun rose higher the men found themselves shedding their outer layers of clothing as they pushed their canoes up the rapids. At times it was necessary to get out of the canoes and wade in the ice-cold water near shore. At any other time they would carry the canoes around these areas, but with the snow on shore so deep, it was almost impossible.

By mid-afternoon the party was back at Allagaskwigamook, and before the sun set each man had returned to his family.

Sugar Maple and the other women and girls had been busy while the men were away. They had laced snowshoes, sewn the winter garments for the hunters, made the wigwams more snug and warm, and built raised beds by placing forked sticks into the ground and then constructing the bed frame over the sticks.

That night the entire band gathered around the fire at Running Bear's wigwam, and the men told of their successful hunt and how they had fared during the storm. Each man talked about his exploits and the story of Walking Deer's three geese was recounted by Hawk-In-Flight.

Returning Thunder had had a very moving experience while the other men were away, and he quietly told them about it.

"On the morning after the night the snows started to come I was helping the women bring wood inside the wigwams. While I was doing this I heard a voice, as if inside my head, that called out my name. I looked up, and there standing beside a tree was a bear. The bear was very thin and weak. It was staggering and struggling to stay on his feet. Then I heard the voice in my head say, 'Returning Thunder of the Bear Clan, kill me that I may be at peace and not do harm to The People.' The bear continued to stand there, and I took my spear and walked up to him. The bear did not move. Again the voice said, 'Kill me.' I drove my spear into his body, and immediately he fell to the ground. He did not struggle, and just before he died I heard the voice for the last time saying thanks. We took the bear far from camp and, as the flesh was inflamed with fever, we buried it as best we could so wolves could not dig it up."

The other men silently gave thanks that their totem had died an honorable death at the hands of one of their clan. Running Bear moved and put his hand on his brother's shoulder for a few seconds in silent appreciation for what he had done.

The children had missed the old uncle's stories while he had been away, so they pleaded with him now to tell them one. Moose-Comes-Running had expected this, and even as he and the others were poling their canoes slowly up the rapids he was recalling a story for the children. Watching the red squirrels in the trees along the river's edge had reminded him of this tale of Gluskabe.

"Do you remember the story I told you about Gluskabe and his twin brother, and how the ugly fish were created?" he asked the young ones.

The children nodded eagerly.

Moose-Comes-Running continued, "Well, as with the fish, Gluskabe's attempts at creating other animals were not always successful, but he kept on until he had fashioned all of the animals and gave them breath. Then Gluskabe made his favorite creature, man. To do this, he gathered a great pile of red clay and then shaped it into a figure that resembled himself. Because he had a great deal of practice at shaping fishes and animals he was now skilled at this kind of work, and he made an excellent model, his best yet. First he made the head to the north, then the arms to the east and west, then the legs to the south. Last, he made the body and connected them all together and breathed on his creation to give it life. This is why that even today we try to sleep with our heads to the north. Where Gluskabe's new creation had been lying, there sprang up a hill of corn. Gluskabe told him that the corn was to be his food.

"Now Gluskabe decided to see how the animals would react to his most wonderful creation, man. He called them all together and said, 'Man,' then watched to see what would happen. The deer and moose ran away. The beaver and otter dove into the water. The birds flew away, and the wolf slunk off into the woods. As for the squirrel, who was then a very large animal, he ran about biting down huge trees, throwing boulders, and making a terrible commotion. Seeing this, Gluskabe called him aside and soothed him and petted him. Each time Gluskabe's hand stroked the squirrel it became smaller until, at last, he was reduced to the size he is today. All of this petting resulted in the curled tail, which the squirrel now has instead of the straight one he had formerly. Gluskabe again said, 'Man,' and again the squirrel reacted as before. But now, instead of biting down trees and throwing boulders, he could only snap off twigs and throw nuts and chatter harmlessly in the branches. This was not exactly what Gluskabe had in mind for the squirrel, but he decided it was good enough, and that is how we came to have the red squirrel we know today."

The children smiled as they thought about the times they had watched the nervous little red squirrel chattering and running about in the trees and the twigs and leaves that always showered down around them.

Sugar Maple motioned for the children to remain seated and then spoke. "While the men were away we were visited by two runners from the Passamaquoddy, who live near the great salt water for much of the year. The young men had been sent out by the Sahgum of the Passamaquoddy to tell all of the Wabnaki tribes of a strange people they had met.

"These people had appeared in a huge canoe on the great salt water after a wind storm. The canoe was larger than any we know. They did not know where it came from, but it carried many strange men with paddles and had one large wing in the center of it. On the front of the canoe was a wooden totem of a strange creature that had never been seen by any of the Wabnaki people.

"The men from the huge canoe had white skins, and their hair was the color of marsh grass in autumn. Their heads and part of their bodies were covered with garments made of a hard substance that shone like the sun shines on the surface of the lake. The weapons they carried were made of the same material, and they were very sharp. The white men spoke a strange language. When the storm that had driven them

ashore was over, they repaired their canoe, got back in, and disappeared from sight. That is all the runners had to say."

No one spoke. All of them were deep in thought about this strange encounter, but they all knew that Sugar Maple would have repeated the story just as she heard it, so there was no point in asking questions, though they all were bursting with them. They would all learn more details when summer came and they were back on the coast near the Passamaquoddies.

Running Bear thought about the story for a long time. He had heard ancient legends of tribes that moved about on the great salt water. But no legend ever mentioned white-skinned men with straw-colored hair. As he put away his stone pipe, Running Bear's inner voice asked, "I wonder if these strange men will change the lives of The People?"

EPILOGUE

The answer to Running Bear's question was not to come for nearly six centuries. Meanwhile the Bear Clan, the other clans of The White Rocks People, and all the other tribes of the Wabnaki continued their lives in much the same way. There were other occasional encounters with the strange white-skinned people, but they were few and far between. Any change wrought by these meetings seemed of little consequence.

Not that the lives of The People went on totally unchanged, for the ability to adapt and change was the most important trait these people possessed to help them survive. Just as the Bear Clan adapted to the loss of hunting area because of fire, so all The People altered their habits to meet each new challenge. Sometimes individuals failed, but overall they were a successful people.

When the change started it came innocently enough at first, because of the Wabnaki's common sense in recognizing the superior tools, clothing, and weapons that the white man had to offer. Nothing in their culture could have prepared them to realize the terrible price they would eventually pay for these things upon which they came to rely within a generation or two. After that, the paradox: They had become a people dependent upon their enemy.

GLOSSARY

GLOSSARY OF NATIVE AMERICAN PLACE NAMES

Allagaskwigamook, "Bark Cabin Lake"—Churchill Lake.
A'pmoojenegamook, "Cross Lake"—Chamberlain Lake.
Kwanosa'gamik, "Loon Lake"—Webster Lake.
Megkwakagamik, "Marsh Pond"—Mud Pond.
Nahmajimskit'egwek, "The dead water extends up into the high land—Smith Brook.

Pataquongamis, "Round Pond"—Round Pond.
Pataweekongomoc, "Burnt Land Lake"—Telos Lake.
Pongokwahemook, "Woodpecker Place"—Eagle Lake.
Sahbimskitegwek, "Empties between two large bodies of water"—Thoroughfare Brook.
Sawanakimoc, "South Lake"—Telos Lake (before fire).
Wallastook, "Beautiful River"—Saint John River.
Wassategnewick, "Place where they spear fish"—East Branch, Penobscot River.

GLOSSARY OF NATIVE AMERICAN WORDS USED

Awe'sus, Bear
Dagwotgawi, Fall (Autumn)
Manitou, Great Spirit, God
Ni'ban, Summer
Ni-datbe, Best Friend
Pabotni, Winter
Pitsonungan, Long Pocket
Sahgum, Chief
Si'gwan, Spring

BIBLIOGRAPHY

Adney, Edwin Tappan and Chapelle, Howard 1., *The Bark Canoes and Skin Boats of North America*, U.S. Government Printing Office, Washington, D.C., 1964.

Anastas, Peter, *Glooskap's Children*, Beacon Press, Boston, Mass., 1973.

Beck, Horace P., *Gluskap the Liar and Other Indian Tales*, The Bond Wheelwright Co., Freeport, Maine, 1966.

Bennett, Dean B., *Maine Dirigo, I Lead*, Down East Books, Camden, Maine, 1980.

Bennett, Dean B., *The Wilderness from Chamberlain Farm*, Shearwater Books, Washington, 2001.

Bonnichsen, Robson & Associates, *Archeological Research at Munsungun Lake*: 1980 Preliminary Technical Report of Activities, Munsungun Lake Paper, No. 1, Institute for Quaternary Studies, Univ. of Maine at Orono, Maine, 1980.

Bull, John and Farrand, John Jr., *The Audubon Society Field Cuide to North American Birds*, Eastern Region, Alfred A. Knopf, Inc., New York, 1977.

Butler, Eva L. and Hadlock, Wendell S., *A Preliminary Survey of the Munsungun-Allagash Waterways*, Bulletin VI11, The Robert Abbe Museum, Bar Harbor, Maine, 1962.

Cook, David and Spiess, Arthur, *Archeology of the Piscataquis Ahwangan*: Preliminary Results, Maine Archeology Society.

DeLorme Publishing Company and Gilpatrick, Gil, *Allagash and St. John Map and Guide*, DeLorme Publishing Company, Freeport, Maine.

Dietz, Lew, *The Allagash*, Holt, Rinehart and Winston, New York, 1968.

Eckstorm, Fannie Hardy, *Old John Neptune and Other Maine Indian Shamans*, A Marsh Island Reprint, Univ. of Maine at Orono, Orono, Maine, 1980.

Eckstorm, Fannie Hardy, *The Handicrafts of the Modern Indians of Maine*, The Robert Abbe Museum, Bar Harbor, Maine, 1980.

Eckstorm, Fannie Hardy, *The Penobscot Man*, Reprint of 1924 edition, New Hampshire Publishing Company, Somersworth, New Hampshire, 1972

Favour, Edith, *First Families; Woodland People of Maine and the Canadian Maritimes*, State of Maine, Dept. of Education and Cultural Services, Augusta, Maine, 1975.

Gagnon, Lana, *Chesuncook Memories*, A collection of articles from the author's column in the "Moosehead Gazette," published by Oscar Gagnon, 1974.

Gilpatrick, Gil, *Building a Strip Canoe, Second Edition, Revised & Expanded*, Fox Chapel Publishing, East Petersburg, Pennsylvania, 2010.

Gilpatrick, Gil, *Building Outdoor Gear*, Fox Chapel Publishing, East Petersburg, Pennsylvania, 1999.

Gilpatrick, Gil, *The Allagash Guide*, Fox Chapel Publishing, Fox Chapel Publishing, East Petersburg, Pennsylvania, 2002.

Gilpatrick, Gil, *The Canoe Guide's Handbook*, DeLorme Publishing Company, Freeport, Maine, 1981.

Gilpatrick, Gil, *The Outdoor Leader's Handbook*, Fox Chapel Publishing, East Petersburg, Pennsylvania, Maine, 2002.

Hamlin, Helen, *Nine Mile Bridge*, W.W. Norton & Co. Inc., New York, 1945.

Hempstead, Alfred Geer, *The Penobscot Boom*, Reprint of a 1931 thesis by the author, 101 Morton Drive, Penny Farms, Florida, 1975.

– BIBLIOGRAPHY –

Holbrook, Stewart H., *Tall Timber,* The MacMillan Co., New York, 1941.

Hubbard, Lucius L., *Woods and Lakes of Maine,* Reprint New Hampshire Publishing Company, Somersworth, N.H., 1971.

Iluber J. Parker, *The Wildest Country,* Appalachian Mountain Club, Boston, 1981.

Jackson, Annette, *My Life in the Maine Woods,* W.W. Norton & Co., Inc., New York, 1954.

Karstad, Aleta, *Wild Habitats,* Charles Scribner's Sons, New York, 1979.

Kephart, George S., *Campfires Rekindled,* Channing Books, Marion, Mass., 1977.

McLeod, John E., *The Northern. The Way I Remember,* Great Northern Paper Company, Millinocket, Maine.

Niss, Bob, *Faces of Maine,* Guy Gannett Publishing Co., Portland, Maine, 1981.

Peterson, Roger Tory, *A Field Cuide to the Birds,* Houghton, Mifflin Company, Boston, 1947.

Pike, Robert E., *Tall Trees, Tough Men,* W.W. Norton & Co., Inc., 1967.

Richardson, Joan, *Wild Edible Plants of New England,* DeLorme Publishing Company, Freeport, Maine, 1981.

Rutherford, Phillip R., *The Dictionary of Maine Place-Names,* The Bond Wheelwright Company, Freeport, Maine, 1970.

Ruthven, John A., and Zimmerman, William, *Top flight Speed Index to Waterfowl of North America,* Moebius Printing Company, Milwaukee, Wisconsin, 1965.

Sanger, David, *Cultural Resource Management in the Dickey-Lincoln School Reservoir, Maine.* A report prepared for the Corps of Engineers under contract with UMO, 1977.

Sanger, David, *Discovering Maine's Archeological Heritage,* Maine Historic Preservation Commission, Augusta, Maine, 1979.

Smith, David C., *A History of Lumbering in Maine 1861-1960*, University of Maine Press, Orono, Maine, 1972.

Speck, Frank G., *Penobscot Man*, Reprint, Octogon Books, 19 Union Square, West, New York, N.Y., 10003, 1970.

The National Geographic Society, *Indians of the Americas*, The National Geographic Society, Washington, D.C., 1955.

Thoreau, Henry David, *The Maine Woods*, Thomas Y. Crowell Company, New York, 1961.

Wells, Robert W., *Daylight in the Swamp*, Doubleday & Company, Inc., Garden City, New York, 1978.

Wilbur, C. Keith, *The New England Indians*, The Globe Pequot Press, Chester, Connecticut, 1978.

Wood, Richard G., *A History of Lumbering in Maine*, 1820-1861, University of Maine Press, Orono, Maine, 1935.

Periodicals

Down East Magazine; April 1956, October 1961, March 1962, May 1962, May 1963, July 1967, September 1967, December 1967, April 1968, March 1973, April 1974, March 1977.

Maine Times; 1/19/73, 8/15/75, 8/22/75.

Rogers, Lore A., *The Telos Cut*, Reprint from "Northern Logger," May 1966.

The Northern; A publication of the Great Northern Paper Company from April 1921 through October 1928.

CONTACT INFORMATION:

By mail: Gil Gilpatrick
P.O. Box 461
Skowhegan, ME 04976

Web site: www.gilgilpatrick.com

Email: gil@gilgilpatrick.com

Telephone: 207-453-6959

Cell: 207-474-1959

Other books by Gil Gilpatrick

The Allagash Guide

What You Need to Know to Canoe
This Famous Maine Waterway

Everything you need to know to take a canoe trip on the Allagash River from expert outdoorsman Gil Gilpatrick.

ISBN: 978-1-56523-488-8 • $11.95 • 104 Pages

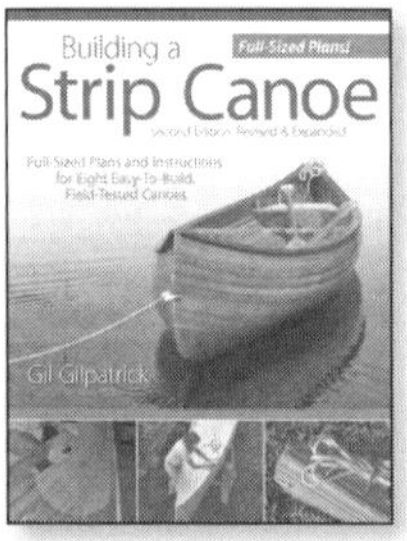

Building a Strip Canoe, Second Edition, Revised & Expanded

Full-Sized Plans and Instructions for
Eight Easy-To-Build, Field-Tested Canoes

Gilpatrick shares his voice of experience in guiding both novice and accomplished woodworkers in building a canoe with easy step-by-step instructions.

ISBN: 978-1-56523-483-3 • $24.95 • 112 Pages

Building Wooden Snowshoes & Snowshoe Furniture

The art of making traditional snowshoes and rustic snowshoe-style furniture is given new life in this beautiful yet practical book.

ISBN: 978-1-56523-485-7 • $19.95 • 160 Pages

Building Outdoor Gear

ISBN: 978-1-56523-484-0

$19.95 • 128 Pages

The Outdoor Leader's Handbook

ISBN: 978-1-56523-486-4

$14.95 • 160 Pages

Look for These Books at Your Local Bookstore

To order direct, call **800-457-9112** or visit *www.FoxChapelPublishing.com*

By mail, please send check or money order + $4.00 per book for S&H to:
Fox Chapel Publishing, 1970 Broad Street, East Petersburg, PA 17520